MOLECULAR BIOTECHNOLOGY

Therapeutic Applications and Strategies

Sunil Maulik

Tripos, Inc.
Redwood City, California

Salil D. Patel

Department of Microbiology and Immunology
Stanford University School of Medicine
Stanford, California

Molecular Biotechnology
Therapeutic Applications and Strategies

Authorized reprint by Wiley India Pvt. Ltd., 4435-36/7, Ansari Road, Daryaganj, New Delhi – 110002.

Reprint: 2014

Printed at: *Radha Offset, Delhi.*

ISBN: 978-81-265-4793-7

MOLECULAR BIOTECHNOLOGY

CONTENTS

Preface ix

Acknowledgments xi

Overview and Introduction xiii

1 Genome-Based Medicine 1

1.1. An Overview of the Human Genome Initiative / 1
Positional Cloning—A Method of Identifying Disease Genes / 6
Inheritance and the Genome / 7
Cloning Disease Genes / 8
Internet Access to Human Genetic Maps / 13
1.2. Strategies for Mapping the Human Genome / 13
Development and Applications of Human Genetic and Physical Maps / 14
Methods for Tracking Disease Genes / 18
Medicinal Applications of a Linkage Map of the Mouse / 19
1.3. The Role of Bioinformatics in Genome-Based Therapy / 20
1.4. Analysis of Human Disease Genes / 24
Huntington's Disease / 24
Cystic Fibrosis / 25
Retinosis Pigmentosa / 27
Alzheimer's Disease / 29
1.5. Therapeutic Genomics in the Private Sector / 31
1.6. Summary / 32
References / 33

2 Human Gene Therapy 37

2.1. An Overview of Molecular Medicine / 37
2.2. Technological Developments in Gene Therapy / 39

2.3. Development of Delivery Systems / 39
Cellular Modalities of Gene Insertion / 39
Other Modalities of Gene Insertion / 41
Applications of Gene Therapy / 45
2.4. Targeting and Control of Gene Expression / 54
2.5. Generation of Mouse Model / 56
2.6. Summary / 63
References / 64

3 Drug Discovery by Combinatorial Chemistry and Molecular Diversity 68

3.1. An Overview of Therapeutic Targets for Drug Discovery / 68
Intracellular Therapeutic Targets / 69
Extracellular Therapeutic Targets / 71
3.2. Combinatorial Chemistry in Drug Development / 72
Pharmacological Definitions of Receptor-Effector Interactions / 73
3.3. Methods for Creating Molecular Diversity / 77
3.4. Principles of Combinatorial Chemistry / 80
Synthesis Strategies for Combinatorial Chemistry / 82
3.5. Molecular Diversity of Proteins / 83
Phage Display of Antibody F(ab) Fragments / 83
Directed Evolution of Proteins of Therapeutic Significance / 88
3.6. Molecular Diversity of Peptides / 90
Soluble Synthetic Peptide Combinatorial Libraries / 90
Chemical Synthesis of Peptides / 91
Drug Discovery Using Oligo N-Substituted Glycines ("Peptoids") / 94
3.7. Molecular Diversity of Oligonucleotides / 97
Randomized Oligonucleotide Libraries / 97
Molecular Diversity of Ribonucleic Acid (RNA) Ligands / 99
3.8. Molecular Diversity of Small Organic Molecules / 101
Screening and Identification of Organic Compounds Through Peptide Tags / 101
Organic Diversity by "Diversomer" Synthesis / 102
3.9. Summary / 103
Methods for Quantitating Molecular Diversity in Combinatorial Libraries / 104
References / 106

4 Protein Engineering and Computer-Assisted Drug Design 109

4.1. An Overview of the Role of Computational Chemistry in Therapeutic Drug Design / 109
4.2. Protein and Antibody Engineering / 110
Knowledge-Based Protein Modeling / 112
Monoclonal Antibody (MAb) Engineering for Immunotherapy / 114
Examples of Therapeutic and Diagnostic Humanized Antibodies / 114
4.3. Peptide and Peptidomimetic Engineering / 120
4.4. Computational Chemistry in Small-Molecule Drug Design / 124
Designing Drugs Without a Target 3D Structure / 124
Computational Aspects of Small Molecule Design / 126
Application of Pharmacophore-Based Drug Design / 130
Structure-Based Drug Design—Use of X-ray and NMR Structures / 138
De novo Drug Design Based on Target 3D Structures / 141
Applications of Structure-Based Drug Design / 142
4.5. Summary / 147
References / 148

5 Re-Engineering the Immune Response 154

5.1. An Overview of the Immune Response / 154
5.2. Organs and Cells of the Immune System / 155
The Arrangement of Antibody and T-cell Receptor Genes / 157
5.3. Trafficking in the Immune System / 159
5.4. Antigen Processing and Presentation by the Immune System / 161
The MHC Genes and Products / 161
Processing Pathways / 164
Structure of MHC Molecules / 165
5.5. Mechanisms of Tolerance Induction / 166
Central Tolerance / 166
Peripheral Tolerance / 166
Immune Regulation/Suppression / 168
The Th1-Th2 Balance / 168
5.6. Specific Immune Modulation / 169
Tolerance Induction with Antigen / 169
Insulin-Dependent Diabetes Mellitus / 170

MHC Blocking Strategies / 172
TCR Blocking Strategies / 175
T-Cell Vaccination / 176
Tilting the Th1-Th2 Balance / 176
5.7. Nonspecific Modes of Immune Intervention / 177
Immunosuppressive Agents Targeting the T Cell / 177
Targeting Accessory Molecules / 178
Targeting Cell Adhesion Molecules / 179
Targeting Cytokines and Related Molecules / 180
5.8. Summary / 180
References / 182

Appendix

Biopharmaceutical Company List / 186

Glossary 191

Index 219

PREFACE

Since the early 1980s the field of recombinant DNA technology has grown enormously, and it has provided a vast quantity of information and tools now being utilized for biotechnology and medicine. This has resulted in a need for instructional and reference materials that reflect the diversity of technologies this field embraces. The idea for this book originated in a class taught by us to biotechnology professionals in the San Francisco Bay Area. Their desire to learn about the methods and technologies utilized at therapeutic biopharmaceutical companies led to our efforts in creating this book. A host of academic textbooks exist that explain the origins, concepts, and scope of biotechnology, yet few describe the application of these concepts to the discovery of novel therapeutic molecules. That was the impetus for this book: to attempt to answer the question of how one converts the knowledge gained from an advance in molecular biology into a commercial therapeutic agent of appropriate efficacy. We have attempted to answer this question by studying current applications of therapeutic biotechnology as directed to novel drug discovery and drug optimization and to the development of novel therapeutic agents, such as genes for gene therapy, and proteins and peptides (that modulate the effect of the immune system) for immunotherapy. Since biotechnology encompasses everything from the production of recombinant proteins to the use of biological molecules as components of nanoscale devices, we have chosen to focus only on applications and strategies to develop therapeutic agents that we feel reflect the state-of-the-art, rather than attempting a comprehensive bookshelf reference. For instance, we have not considered research topics such as gene regulation and transcription or the elucidation of signal transduction pathways, concentrating instead on the targets these types of studies provide for the drug developer.

In each chapter of this book, we have attempted to cover the trends, applications, and strategies that we believe show the most promise for therapeutic intervention. The emphasis is on the interdisciplinary approach needed to interface research with development and create commercial therapeutic molecules. Appendixes and a glossary that provide explanations of some of the basic concepts of molecular biotechnology are included. We hope that the multifaceted nature of the book, reflecting our own interests in biochemistry, biophysics, chemistry, and computer science, as well as our perspectives from both academia and industry, will provide the reader with a new and unique viewpoint of this fast moving field. We welcome the remarks of readers on the content and design of the book via the publisher.

SUNIL MAULIK
SALIL D. PATEL

ACKNOWLEDGMENTS

We would like to acknowledge our students for first suggesting our undertaking this project, our editors, Susan King and Colette Bean, for their assistance in the development of the book, and Kenneth Hobbs, Jr. for his assistance with the figures. We also wish to thank the reviewers for their many useful comments which improved the final version of the manuscript immeasurably. S. M. wishes to thank Mike Sullivan and Tripos, Inc. for assistance with the computer graphics images, and Drs. Sunil Chada, Peter McWilliams, Pradip Rathod, and David Mosenkis for their critical reading of the manuscript. Thanks also to Sonali Maulik for her comments on some of the figures.

OVERVIEW AND INTRODUCTION

Recombinant DNA technology, commonly referred to as genetic engineering, has revolutionized our ability to isolate, manipulate, and express genes (and thus proteins), virtually at will. The first medicinal benefits of this have been the expression and isolation of numerous therapeutic proteins (including antibodies). Some of these genetically engineered proteins approved for therapeutic use are listed in Tables I and II. In most cases these therapeutic proteins provide replacement therapy for defective, inactive, or absent proteins in patients with well-characterized biochemical defects. The use of insulin to treat diabetes, or adenine deaminase to treat "boy in the bubble syndrome," or growth hormone to treat dwarfism are examples of this type of protein therapy. In other cases the therapeutic proteins may act to stimulate parts of or all of the immune system to generate a favorable response. The use of colony-stimulating factors to stimulate platelet and neutrophil (blood cells) production in neutropenic patients, and interferons and interleukins for stimulating an anticancer response, represent such cases.

Many therapeutic proteins are now expressed in bacterial, yeast, insect, or mammalian cell systems, which can now be readily manipulated using the techniques of genetic engineering. In some of these systems, large bioscale production may be difficult, so proteins are being expressed in transgenic animals and plants that act as expression vats to make the proteins—a process known as "pharming." Examples of proteins made by this approach include lactoferrin and protein C, made in cows and used in infant formula and blood clotting (GenPharm); the cystic fibrosis transmembrane receptor (CFTR) protein expressed in goats as a therapy for cystic fibrosis (Genzyme); alpha-1 antitrypsin in sheep as a therapy for emphysema (Pharmaceutical Proteins Ltd.); serum albumin in potatoes as a blood expander (Mogen SA); enkephalin in rape plants as a painkiller (Plant Genetic Systems); and melanin in tobacco plants as a sunscreen (Biosource Genetics). Pharming with transgenic animals and plants may also allow the industrial-scale pro-

Table I Novel therapeutic products from biotechnology

Year	Product	Disease	Company
1982	Humulin (synthetic insulin)	Type I diabetes	Genentech, Inc.
1985	Protropin (human growth hormone)	Growth hormone deficiency	Genentech, Inc.
1986	Intron A (interferon-alpha)	Hairy cell leukemia	Biogen, Inc. and Schering-Plough, Inc.
	Roferon A (interferon-alpha)	Hairy cell leukemia	Genentech, Inc. and Roche, Inc.
	Orthoclone OKT3 (monoclonal antibody)	Kidney transplant rejection	Ortho Biotech, Inc. (Johnson & Johnson)
	Recombivax AV (synthetic vaccine)	Hepatitis B	Chiron Corp. and Merck and Co.
1987	Humatrope (human growth hormone)	Growth hormone deficiency	Biogen, Inc., and Eli Lilly, Inc.
	Activase (tissue plasminogen activator)	Myocardial infarction	Genentech, Inc.
1988	Intron A	Genital warts	Biogen, Inc., and Schering-Plough, Inc.
	Roferon A	Genital warts	Genentech, Inc. and Roche, Inc.
1989	Epogen (erythropoietin)	Renal anemia	Amgen, Inc.
	Engerix-B	Hepatitis B	Smith Kline Beecham, Inc.
	Alferon N	Genital warts	Interferon Sciences, Inc.
1990	CytoGam	Cytomegalovirus infection	
	Activase	Pulmonary embolism	Genentech, Inc.
	AlphaNine	Hemophilia B	
	ProCrit (erythropoietin alfa)	AIDS/AZT anemia	Ortho Biotech, Inc.
	Actimmune (interferon-gamma)	Chronic granulomatous disease	Genentech, Inc.
	Adagen (adenosine deaminase)	Severe combined immunodeficiency (SCID)	Enzon, Inc.
1991	Intron A	Hepatitis C	Biogen, Inc., and Schering-Plough, Inc.
	Neupogen (colony-stimulating factor)	Neutropenia from chemotherapy	Amgen, Inc.
	Leukine (GM-CSF)[a]	Neutrophil recovery from bone-marrow transplants	Immunex, Inc.
	Ceredase (alglucerase)	Gaucher's disease	Genzyme Corp.
	Provisc	Ophthalmic surgery	Chiron Corp.

Table I (***Continued***)

Year	Product	Disease	Company
1982	Proleukin (interleukin-2)	Renal cell cancer	Chiron Corp.
	Alpha Nine SC	Factor IX viral infection	
	Intron A	Hepatitis B	Biogen Inc. and Schering-Plough, Inc.
	Recombinate (factor VIII)	Hemophilia A	Genetics Institute, Inc. and Baxter Corp.
	OncoScint OV103 (monoclonal antibody complex)	Ovarian cancer imaging	Cytogen Corp.
	OncoScint CR103 (monoclonal antibody complex)	Colorectal cancer imaging	Cytogen Corp.
1993	KoGENate (antihemophiliac factor)	Hemophilia A	Miles, Inc.
	Procrit	Anemia from chemotherapy	Ortho Biotech
	Orthoclone OKT3	Heart and liver transplant rejection	Ortho Biotech (Johnson & Johnson)
	Betaseron (interferon-beta)	Multiple sclerosis	Berlex Labs., Inc. and Chiron Corp.
	Neutropin	Short stature (associated with chronic renal insufficiencies)	Genentech, Inc.
	Pulmozyme (DNase)	Cystic fibrosis	Genentech, Inc.
1994	Oncospar (pegaspargase)	Acute lymphocytic leukemia	Enzon, Inc., and Rhone-Poulenc Rorer
	Neutropin	Growth hormone deficiency	Genentech, Inc.
	Cerezyme (imiglucerase)	Gaucher's disease	Genzyme, Inc.
	Neupogen	Bone-marrow transplants	Amgen Corp.
	Albunex	Diagnosis of heart disease	Molecular Biosystems (Mallinckrodt)

[a] GM-CSF stands for granulocyte macrophage–colony-stimulating factor.

duction of complex proteins that could not be expressed in bacteria or yeast, such as human collagen, which has important uses in skin healing, tissue transplantation, and dermatology.

The genetic engineering revolution has also provided us with a mass of information regarding the mechanisms of cancer and the development and

Table II Therapeutic products in phase III clinical trials 1995

Product	Company	Disease
Antisense		
ISIS 2922	Isis Pharmaceuticals, Inc.	Inflammatory diseases
Clotting factors		
NovoSeven (recombinant factor VIIa)	Novo Nordisk	Hemophilia A & B
Colony-Stimulating Factors		
Leukine (GM-CSF)	Immunex	Neonatal sepsis
Neupogen (recombinant G-CSF)	Amgen, Inc.	Hematologic disorders
Pixykine (PIXY321)	Immunex	Neutropenia and platelet deficiency
Dismutases		
DISMUTEC PEG-superoxide dismutase	Enzon, Inc.	Closed head injury
Erythropoietin		
PROCRIT Erythropoietin alpha	Ortho Biotech, Inc.	Anemia, blood adjuvant
Growth Factors		
Igef (recombinant insulin-like growth factor)	Pharmacia	Growth hormone deficiency
Myotrophin (recombinant insulin-like growth factor)	Cephalon, Inc.	Amyotrophic lateral sclerosis
PDGF Platelet-derived growth factor	Chiron/R. W. Johnson	Wound healing
Human Growth Hormones		
Serostim (recombinant somatotropin)	Serono Labs.	AIDS-associated weight loss
Interferons		
Actimmune Interferon-gamma 1b	Genentech, Inc.	Renal cell carcinoma
Betaseron (recombinant interferon beta-1b)	Berlex Labs., Inc.	Multiple sclerosis
Consensus interferon	Amgen, Inc.	Hepatitis C
Intron A (recombinant interferon alpha-2b)	Schering-Plough, Inc.	Miscellaneous cancers
Beta-interferon 1a	Biogen, Inc.	Multiple sclerosis
Roferon-A	Hoffmann-LaRoche	Hepatitis C
Interleukins		
Proleukin Interleukin-2	Chiron Corp.	Cancer and HIV infection
Interleukin-3	Sandoz Corp.	Platelet engraftment
Interleukin-6	Sandoz Corp.	Platelet engraftment
Monoclonal Antibodies		
anti-HER-2 humanized mAb	Genentech, Inc.	Breast cancer
anti-tumor necrosis factor mAb	Chiron Corp.	Sepsis

Table II (***Continued***)

Product	Company	Disease
E5 mAb	Pfizer/Xoma	Gram-negative sepsis
IDEC-C2B8 pan-B mAb	IDEC Pharm.	Non-Hodgkin's B-cell lymphoma
enlimomab (anti-ICAM-1 mAb)	Boehringer Ingelheim	Kidney transplants, stroke
Oncolym mAb	Alpha Therapeutics	Non-Hodgkin's B-cell lymphoma
ONCOLYSINB (anti-b4-blocked ricin)	ImmunoGen	B-cell leukemia/lymphoma
ReoPro abcixi-mAb	Centorcor/Eli Lilly	Restinosis, angina
SMART (anti-tac mAb)	Protein Design Labs.	Graft vs. host disease
Zenapax daclixi-mAb	Hoffmann-LaRoche	Graft vs. host disease
Tissue Plasminogen Activators		
Activase (recombinant tissue plasminogen activator, t-PA)	Genentech, Inc.	Ischemic stroke
Vaccines		
DTaP	Biocine/Chiron	Diphtheria, tetanus, pertussis
Herpes simplex	SmithKline Beecham	Herpes
Herpes vaccine	Biocine/Chiron	Herpes, genital herpes
HIV vaccine	Immune Response Corp.	Asymptomatic HIV infection
Lyme borreliosis vaccine	Connaught Labs.	Lyme disease
Lyme vaccine	SmithKline Beecham	Lyme disease
Melacine	Ribi Immunochem	Stage III, stage IV melanoma
VaxSyn HIV-1	MicroGeneSys	AIDS
Miscellaneous		
Auriculin anaritide (atrial natriuretic peptide)	Scios Nova & Genetech, Inc	Acute renal failure
Ceprate SC	CellPro, Inc.	Cell therapy
Hirudin	Ciba Pharm.	Acute coronary syndrome
IL-2 fusion toxin (DAB^{389}-IL2)	Seragen, Inc.	Cutaneous T-cell lymphoma
lhADI (human leutenizing hormone)	Serono Labs.	Follicular development
OP-1 (osteogenic protein-1)	Creative Biomolecules	Non-union fractures
Pulmozyme (DNase alpha)	Genentech, Inc.	Chronic obstructive pulmonary disease
T4N5 liposome lotion (T4 endonuclease V)	Applied Genetics	Xeroderma pigmentosa
Thyrogen (human thyroid-stimulating hormone)	Genzyme, Inc.	Recurrent thyroid cancer
VISTIDE (cidofovir)	Gilead Sciences, Inc.	CMV retinitis in AIDS patients

Source: Genetic Engineering News, August 1995.

workings of both the nervous and immune systems. Therapy for aberrant neurological, immunological, or cellular function can now be pinpointed in those cells that modulate each of these functions. In the case of cancer, a number of factors that regulate the cell cycle have been identified. Classes of therapeutic targets in the cell are summarized in Fig. A.

Although the list in Table I is an impressive tribute to the power of recombinant technology, it is only one method of disease prevention by molecular targeting. In many diseases the defect (target enzyme, protein, or receptor) is unknown, as is the case for all genetic diseases in which a disease gene has not been identified, and the consequences of the disease may be too widespread to alleviate by protein or small-molecule therapy. Examples of such diseases include Huntington's chorea, amylotropic lateral sclerosis (Lou Gehrig's disease), and cystic fibrosis. In such cases the best therapy would be to define the genetic defect in order to correct it, if possible, by "cutting the disease off at its roots." The Human Genome Project aims to delineate the entire human genome. The effort to characterize all 100,000 or more genes of the human genome (described in Chapter 1) will identify many therapeutic targets and disease genes. Gene therapy, the replacement of an aberrant gene with a functional one, may then be utilized to treat some of the above-mentioned diseases. This approach is described in Chapter 2.

Many other diseases are better targeted by conventional "small-molecule" drugs. Two revolutionary techniques in small-molecule drug discovery are combinatorial chemistry combined with high-throughput screening, which enables vast numbers of new and unique potential drug compounds to be tested, and computer-assisted drug design, which allows the systematic exploration of the physicochemical properties of a candidate drug molecule and its possible modes of interaction with a target receptor. This procedure works best when the target has been identified, isolated, and its three-dimensional structure elucidated (a process known as *structure-based drug design*). The discovery and design of new drugs is covered in Chapters 3 and 4. Other diseases, such as the autoimmune diseases, are the result of an aberrant immune response and not simply a particular protein that is defective or lacking. Here an entire network or system has failed to act in an appropriate manner, resulting in disease. Diabetes, rheumatoid arthritis (RA), and multiple sclerosis (MS) are examples of autoimmune diseases. The different therapeutic strategies currently being developed to treat a wide array of autoimmune and immunodeficiency diseases for which conventional therapies are unavailable are reviewed in Chapter 5. A listing of therapeutics (in phase III trials) developed by these means is given in Table II.

Throughout the book we have attempted to tie together the diverse threads that link the different therapeutic approaches to disease prevention exempli-

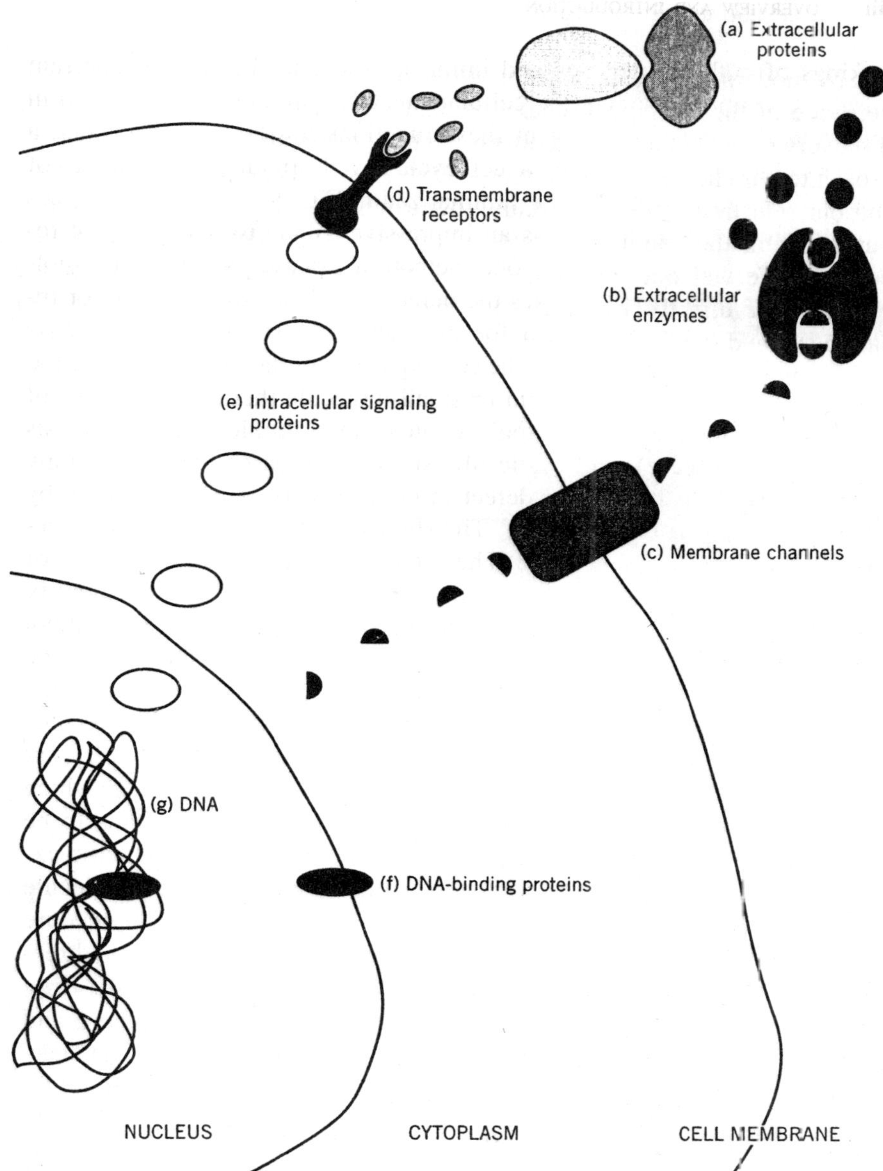

Figure A Therapeutic targets in the living cell. Suitable therapeutic targets being identified by molecular biology include extracellular agents such as proteins (*A*) and enzymes (*B*) secreted by cells that mediate cellular function in disease; cellular channels (*C*) and receptors (*D*) that allow the passage of molecules that signal cells to follow particular pathways that may underlie a disease pathology; intracellular "second messengers" intracellular receptors—molecules within the cell that convey a "message" from an extracellular agent from the cell membrane to the nucleus (*E*) and ultimately regulate the expression of genes that alter the cells behavior; and targets within the nucleus itself such as the transcription and replication machinery of cells (*F*), and the DNA packaged into chromosomes (*G*).

fied by modern molecular biotechnology. In particular, we consider the convergence of the biological, chemical, and informational approaches to the discovery of novel targets and drugs. We have taken an entirely selective approach to our choices of applications, strategies, and companies, and as a result our selections reflect a certain arbitrariness. We look forward to contributions from the community that will address these deficiencies in future editions. We welcome the reactions of our readers, either by way of the publisher or directly by electronic mail to sunilm1@aol.com (S.M.) or sallyp@leland.stanford.edu (S.P.).

MOLECULAR BIOTECHNOLOGY

1

GENOME-BASED MEDICINE

1.1. AN OVERVIEW OF THE HUMAN GENOME INITIATIVE

The Human Genome Initiative (HGI) is a worldwide project that plans to decipher, at the level of each base of the DNA molecule, all the information stored in the genome of humans and of a few other select ("model") organisms. Deciphering and cataloging this information is an enormous project, yet significant advances have already been made. The human genome is estimated to be composed of about 100,000 genes, corresponding to 3×10^9 base pairs (bp) of DNA. Only a fraction of these genes are expressed in any particular tissue, however. The HGI is already having a significant effect on therapeutic medicine, allowing the early diagnosis and intervention treatment of numerous diseases by utilizing DNA-based strategies. This chapter reviews the initiative from the viewpoint of the molecular biotechnologist attempting to understand the significance of the project to their own research and development for the discovery of new therapeutic drugs. Motivations for the genome projects of various organisms vary according to the organism studied, as does the type of data collected and their therapeutic significance.

- *DNA and protein sequence data.* Help our understanding of development and function of the organism in terms of nucleic acid and protein and their fundamental interactions in the cell (described in more detail in this chapter).
- *Gene identification and genetic mapping.* Enable diagnosis and alleviation of symptoms of human genetic disease (more on this later in this chapter and in Chapter 2).

- *DNA and protein sequences and structures.* Provide for the development of new or improved products and therapies by way of protein engineering, biotechnology, and medicinal chemistry (described more fully in Chapters 3 and 4).
- *Complex genetic loci, gene rearrangements, and genetic polymorphisms.* Lead to design of highly specific therapeutics based on patterns of genetic variation in strains, tissues, cellular systems, and organisms (described more fully in Chapter 5).
- *Genetic mapping and qualitative locus studies.* Use animal and plant breeding to improve products such as meat quality or seed oil yield, topics beyond the scope of this book.

When first proposed in the late 1980s, the plan to completely map and sequence every human gene (as well as those of some other organisms) seemed ambitious even to the most optimistic protagonists. Even after an official sanctioning on October 1, 1990, few expected the extent of progress that has been achieved so far. In fact, thanks to rapid technological developments, the first goal, to map the entire human genome, has essentially been completed. (Refinements to the map are expected to occupy researchers for many years to come, however.) This intention is analogous to completing a map of a new country delineating every town, city, and state. The second goal, to sequence the entire human genome (in the analogy above, listing every single address in the map of the new country), is progressing apace, although most scientists believe significant but not insurmountable technological improvements will be needed to complete the sequencing arm of the project by the year 2005. The genetic maps, the coarsest of the possible genome maps, utilize genetic loci that are polymorphic in the human population to determine the inheritance pattern of these loci in families. Nearby loci show similar inheritance patterns, allowing their proximity to be inferred. Genetic linkage is reliable over distances of about 30 Mega (M) bases, given the recombination rate (crossing over) of human chromosomes. Radiation hybrid (RH) maps, utilizing hybrid cell lines containing many large chromosomal fragments provide the next level of mapping detail. The fragments, induced by radiation damage, are PCR-amplified to identify those hybrids that have retained a given locus. Nearby loci tend to show similar retention patterns, again allowing proximity to be inferred.

RH linkage can be detected up to distances of about 10 Mbases, given the average fragment size of the RH panel used. The finest detail of mapping achieved so far is sequence-tagged site mapping, (Fig. 1.1) whereby yeast artificial chromosome (YAC) libraries are screened by PCR to identify all clones containing a given locus. Identification is obtained by sequencing a small fragment of the clone sufficient to uniquely identify it by PCR am-

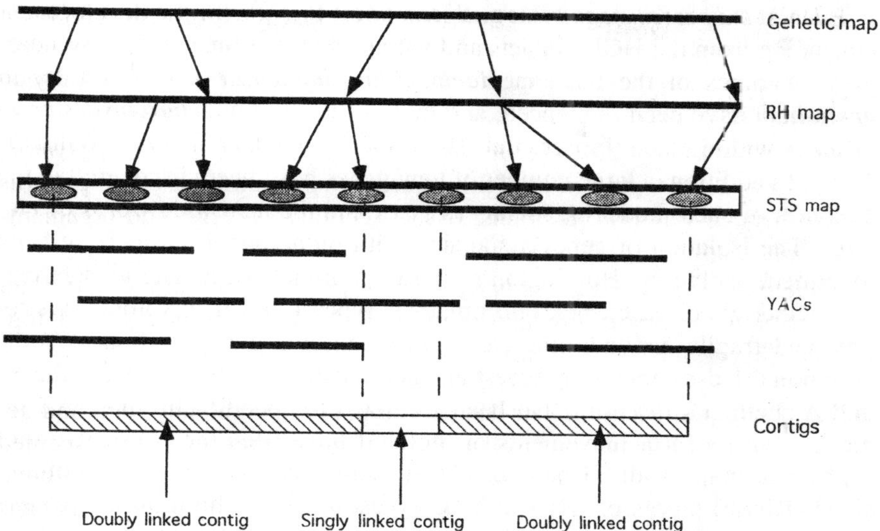

Figure 1.1. Genetic and physical mapping of the human genome. A schematic diagram of the genetic, radiation hybrid (RH) and sequence-tagged site (STS) map of the human genome is displayed. Loci that are genetically mapped are shown as arrows to their positions in the RH map, and the STSs to which these loci map are shown as shaded ovals. Yeast artificial chromosomes (YACs) containing the STSs are shown as overlapping solid lines. STSs may fall into doubly linked contigs (striped rectangles) or into singly linked contigs (clear rectangle). Single linkage is not generally reliable for connecting arbitrary doubly linked contigs except when the doubly linked contigs are known to be adjacent on the genetic or RH map (after Hudson et al. 1995).

plification. Nearby loci tend to be present in many of the clones, again allowing proximity to be inferred. STS-content linkage can be detected over distances up to 1 Mbase, given the average insert size of the YAC library used. Goals delineated by the director of the HGI program in the United States, Francis Collins, include refinement of the genetic map, creation of a 100kb sequence-tagged site (STS), and increase DNA sequencing to a level of at least 50 Mbases/year at a cost of no more than $0.40/base (Collins and Gallas 1993). Sequencing is now typically achieved using random sampling (called "shotgun" sequencing) of clones the size of cosmids or larger, implemented on commercial four-color fluorescence sequencing instruments. Cloning the human genome into yeast or bacterial artificial chromosomes (BACs), the latter which can hold up to 350 kbases of DNA, may ultimately allow the entire human genome to be sequenced by shotgun sequencing strategies, provided that the assembly software is powerful enough to correctly combine the hundreds of thousands of overlapping contigs that would be generated.

Even at this early stage biologists can point to some impressive benefits emanating from the HGI (Guyer and Collins 1995; Collins 1995). The complete genomes of the bacteria *Haemophilus influenzae* and *Mycoplasma genitalium* have been sequenced, and that of the yeast *Saccharomyces cerevisiae* is within reach (Fraser et al. 1995; Fleischmann et al. 1995; Williams 1995). In addition, a large number of new genes have been discovered in the genomes of such model organisms as yeast and the nematode worm *C. elegans.* The isolation of genes associated with numerous diseases have been identified, including Huntington's disease, amylotrophic lateral sclerosis (Lou Gherig's disease), neurofibromatosis types I and II, myotonic dystrophy, and fragile X syndrome. Genes that confer a predisposition to some common diseases, including breast and colon cancer, hypertension, diabetes, and Alzheimer's disease, have been localized to specific chromosomal regions. Two recent achievements of the HGI have been the construction of a physical map of all 23 pairs of chromosomes consisting of 40 kilobase-sized (Kbase) pieces of DNA in YAC (yeast artificial chromosome) clones by Daniel Cohen and the CEPH group of France (Murray et al. 1994), and the recent publication of a 15,000 sequence-tagged site map at ~200 kbase spacing by an international group led by the Whitehead Institute (Hudson et al. 1995). The physical map will enable researchers to home in and pinpoint disease genes much more rapidly then ever before, and the availability of the clones will save several years of effort in isolating and sequencing specific genes. The STS map provides the impetus now to begin large-scale sequencing projects that will "stitch together" the entire human genome sequence from the identified sequence fragments. A 30,000-STS map at a resolution of about 100 kbases is expected within another two years.

Before gene-based therapies can be attempted, the diseases must be diagnosed at the molecular level. The Genome project is already presenting examples of how this can occur; some of these are described below. For example, the discovery of the instability of repetitive DNA sequences (sometimes called "microsatellites" because of their banding appearance in ultracentrifuge tubes) has been associated with numerous types of cancers and other heritable diseases. These microsatellites are typically one- to six-nucleotide motifs tandemly repeated numerous times throughout portions of the genome. The loss or gain of these DNA microsatellites can cause genetic defects, since errors occurring through strand slippage during DNA replication are unable to be corrected as a result of post-replication heteroduplex repair. These repeats are unstable in colon cancer cells, and a gene on chromosome 2 has been implicated in this instability. Expansion of trinucleotide repeats from a few to hundreds or thousands of copies has been associated with fragile X syndrome, Kennedy's disease, myotonic dystrophy, Huntington's disease, and spinocerebrellar ataxia type 1. Some diseases are characterized more by satellite expansions than contractions, leading to speculation that there is a heteroduplex repair mechanism for misaligned or unaligned ("slippery") DNA. Because cancer is a multistage disease requiring several mutations, and it has been speculated that cancer cells possess a distinct "mutator" phenotype, the observation of microsatellite instability in human colon cancer cells suggests that there may be different repair mechanisms for point mutations than for microsatellites.

Technologies developed for the HGI have powerful implications for the development of therapeutic drugs and for understanding their effects in model organisms. Mapping genes chromosome-by-chromosome is now possible using YAC clones. Improved vector systems for cloning large DNA fragments have also been developed, and better experimental strategies and computational methods for assembling clones into contigs (stretches of DNA that together can be connected into a single contiguous gene sequence) by searching for overlaps now exist. High-throughput cDNA sequencing (the sequencing of just those genes known to be expressed into mature mRNA transcripts in the cell) has been developed in a number of laboratories. Bioinformatics, the science of studying biological information at the molecular and cellular level by computer, has become a science in its own right. Some applications of these technologies are already being exploited by genome "boutiques" as well as large pharmaceutical houses as means for identifying new drug targets and therapeutic molecules. Another therapeutic benefit of generating high-resolution linkage maps of model organisms is that they can serve as models for many human diseases. This is true of the mouse, pig, sheep, and cow. High-resolution linkage maps and the use of YAC vectors now allow cloning of just about any mutation known to be implicated in a disease state (these topics are described in more detail below).

One of the great benefits of generating high-resolution genetic maps of different model organisms is as a starting point for positional cloning studies—that is, chromosomal walking from nearby genetic markers (see the sidebar on positional cloning). Until recently this approach was largely impractical because of the paucity of molecular markers in the genetic maps of different organisms and the absence of efficient "walking" methods and large insert libraries. However, given the great increase in marker density of simple sequence-length polymorphism (SSLP) markers, and the ability to clone large-insert YAC libraries of the mouse genome, it is now possible to positionally clone virtually any mouse mutation. As physical maps of the mouse (and other organism) genomes become available and techniques for identifying genes in large blocks of genomic DNA improve, positional cloning will become the strategy of choice in cloning out any mammalian gene desired.

POSITIONAL CLONING—A METHOD OF IDENTIFYING DISEASE GENES

Positional cloning of disease genes requires the availability of a large family with many afflicted individuals; this is straightforward for a disease as prevalent as cystic fibrosis (CF). The cystic fibrosis defect was mapped utilizing linkage analysis. Here a comparison is made within a family for the co-inheritance of markers of known chromosomal location with the mutant gene. The human genome is virtually identical between individuals, with differences (or polymorphisms) typically limited to either a single base change or to a more elaborate variation in the number of repeats of DNA base doublet and triplet motifs, known as microsatellites or variable numbers of tandem repeats (VNTRs) that are found to vary from individual to individual. Although these differences are generally not in coding regions and may have no functional significance, they are extremely important for linkage analysis since they act as markers. During the development of the sperm and ova, homologous chromosomes exchange segments in a process called recombination. The closer a marker is to a given gene, the higher the chance that there will be no recombination between them and that they will be co-inherited despite this genetic reassortment. Co-inheritance thus suggests that the marker and the mutant gene are physically juxtaposed on the chromosome. Linkage analysis was successfully utilized to map the CF gene to its location on chromosome 7. The availability of an increasing number of markers and the development of novel techniques for screening makes linkage analysis valuable. The markers provide the starting point to utilize recombinant DNA cloning methods to walk or jump to the gene of interest. The walks and jumps help to identify other markers that are closer and therefore more tightly linked to the defective gene. The availability of markers that saturate the genome from the CEPH collaboration will provide the higher-resolution map needed in order to identify a desired gene with less random walking or jumping. The final identification of a gene, however, still relies on analyses of overlapping sequence fragments from shotgun sequencing in the finely mapped zone. This might include computational searches for open reading frames and other motifs such as CpG islands that are normally present in structural genes. Absolute identification of the target gene, can only be made by the demonstration of mutational differences in the gene sequences between patients and control subjects.

The human genome is composed of at least 100,000 genes, and any defects within them, such as a single base loss or change (point mutation) or loss of multiple bases (deletion) may cause a number of diseases (currently estimated at over 4000). In some cases deletions or interruptions in genes are the result of chromosomal translocations where large fragments are exchanged between chromosomes. Genes at a chromosomal breakpoint may be inactivated by physical disruption. Each chromosome has a particular size and characteristic banding pattern when the cells are stained with certain dyes, and so such translocations are readily visualized by cytogeneticists. Mutations can change the amino acid sequence of a protein, cause a frame shift, affect RNA processing or stability, or introduce stop codons that affect transcription. Any of these mutational events will potentially lead to a change or loss of function of a protein and thus potentiate a cellular effect that may be deleterious to the individual.

Some disorders are a result of defects in multiple genes of an individual and are termed *polygenic.* Many of these same diseases are also multifactorial in that a complex interaction between these multiple defects and the environment are required to elicit disease. Examples of multifactorial diseases include type 1 diabetes, schizophrenia, and hypertension. These diseases are not directly amenable to gene-based therapy (Chapter 2) and other approaches must be tried (see Chapter 5 for a description of therapeutic approaches to diabetes intervention). Autosomal dominant diseases are also less amenable to gene therapy, since a normal allele already exists in these diseases and specific modification of the abnormal gene only is required. The first candidates for gene therapy were monogenic, autosomal, recessive, or X-linked diseases. Many diseases are in fact a result of multiple defects in the same gene. The thalassemias are a result of defective beta globin genes; different mutations that affect almost all of the above mentioned processes are found in different individuals. In cystic fibrosis 70% of individuals (in northern Europe; figures vary among different populations) affected carry a 3-bp deletion which results in the loss of a key Phenylalanine amino acid. However, the other 30% of cases are a result of some 60 different mutations. These observations make the development of diagnostic assays and screening programs difficult. They do not, however, affect the design of gene therapy protocols, since only the introduction of a normal gene is required.

Inheritance and the Genome

The human genome consists of 22 pairs of so-called autosomal and one pair of sex chromosomes. Each pair consists of one chromosome inherited from the father and one from the mother. The sex chromosomes are different from the autosomes; a male carries an X (inherited from the mother) and a Y (which is inherited from the father and carries the male-determining genes) chromosome and a female has two X chromosomes (derived from each parent). Cystic fibrosis is an example of an autosomal recessive disease; that is, individuals must be homozygous for the cystic fibrosis defect to be affected. Those that are heterozygous will not be afflicted but may pass the defect to their children, who are referred to as *carriers.* The children of carriers thus also have the potential to get the disease. Huntington's disease is an example of an autosomal dominant disorder; a single copy of the mutant gene is sufficient to cause disease. Sex-linked diseases such as color blindness, hemophilia, and muscular dystrophy affect males and females differently because the genes responsible are located on the sex chromosomes. Most typical X-linked diseases are recessive. In these cases mothers pass a defective gene on a X chromosome to their children. Only males are affected, however, since females have a normal gene on their second X chromosome that compensates for or masks the defect. Daughters will, however, become carriers. These modes of inheritance are illustrated in Fig. 1.2 (see page 10).

Cloning Disease Genes

Identification of the gene responsible for a genetic defect and its normal counterpart are the first step toward gene therapy. In 1911 the first human gene was mapped to a specific chromosome. Since color blindness only afflicts males (and is therefore a sex-linked disease), it was mapped to the X chromosome. Other genes, such as those for hemophilia, were also mapped to the X chromosome because of their similar sex-linked mode of inheritance. In the late 1960s geneticists were restricted to using hybrids of somatic cells only to map genes to autosomes. The emergence of cloning technology, however, has led to rapid advancements in human genetics. The isolation of genes for numerous characterized proteins had become routine by the mid 1980s. Nonetheless, the cloning of specific disease causing genes remained slow because the biochemical defect in most of them was as yet unidentified and thus no starting point for cloning existed by the standard technique of screening libraries. Not surprisingly, the first human disease genes were isolated for diseases where the biochemical defect was known. For example, Lesch-Nyhan syndrome is caused by an abnormal gene for hypoxanthine phosphoribosyltransferase (HPRT). The mouse gene for HPRT was isolated utilizing knowledge of the "salvage" pathway that HPRT participates in to metabolically select cells that were overexpressing the enzyme. The human homolog was then isolated by screening a fetal liver cDNA library. In general, the isolation of genes for diseases where the biochemical defect was unknown was achieved by a combination of fortuitous chromosomal abnormalities along with serendipity. For instance, Duchenne's muscular dystrophy (DMD) is an X-linked disorder causing progressive muscle degeneration in young boys. However, some women were observed who also appeared to have DMD. This was explained by the fact that these women have a chromosomal translocation where part of their X chromosome has exchanged with an autosomal chromosome fragment. Although the autosomal fragment varied in these patients, the X chromosome breakpoint was always the same. In addition inactivation of one of the X chromosomes, which occurs randomly from cell to cell in normal women, occurred selectively in the normal X chromosome. Thus these women carried a normal putative DMD gene on one normal X chromosome and a homologous gene on the other active X chromosome, which was postulated to be mutated by the translocation event. It was proposed that the DMD gene mapped to the chromosomal breakpoint. A separate study came to the same conclusion based on the observation that a young boy showing signs of DMD had a X chromosomal deletion at the same location. Further analysis in both cases allowed the isolation of DNA probes that would be eventually used to isolate the DMD gene. The breakpoint was found to be in a region known to code for ribosomal RNA; this provided the start point for a detailed search of that region. In the latter case a subtraction cloning strategy was used to isolate DNA probes specific to the deleted region. Chromosomal abnormalities have also been found to be associated with the genes for retinoblastoma, Wilm's tumor, and chronic granulomatous disease, among others.

New strategies have been developed to isolate genes responsible for diseases where the biochemical defect is unknown and no abnormality-related clues as to the chromosomal location of the gene exist. The prototype for this positional cloning or reverse genetics approach was first demonstrated with the successful cloning of the gene defective in cystic fibrosis in 1989.

A number of "genomics" startup biotechnology companies are already exploiting human genome information in medical and industrial applications. The use of hybridization subtraction techniques between expression maps of cells allows "snapshots" of cellular processes in normal and diseased cells to be compared. Then, by dissecting the expression of a suite of disease genes, the proteins they express can be determined, revealing the underlying mechanism in the cell of the disease genes on the disease. (Subtraction techniques are described in more detail below.) This procedure has enabled the discovery of mRNA (cDNA) molecules that are involved in specialized gene expression. The use of expressed sequence tags (ESTs), partial cDNA stretches that can be amplified by the PCR method, has revealed gene expression from specific tissues (see Fig. 1.3 on page 12). The advantage of EST sequencing is that only those genes expressed in a particular tissue type are detected (by

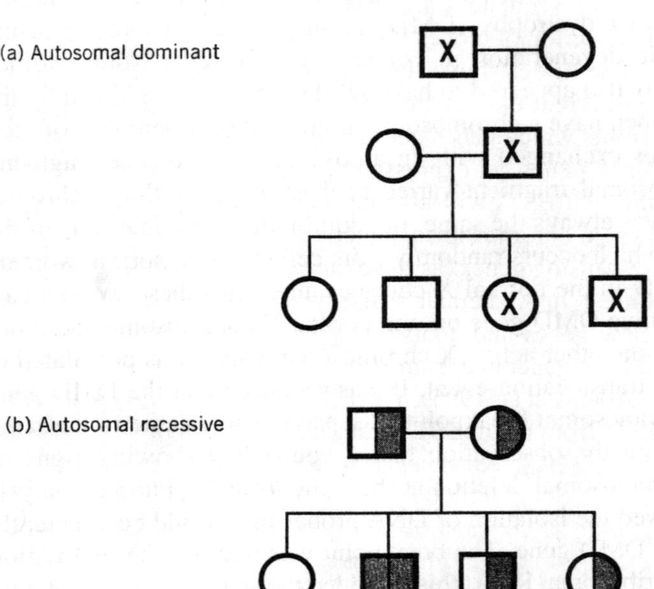

(a) Autosomal dominant
X
X
X
X
(b) Autosomal recessive
Afflicted
Carriers

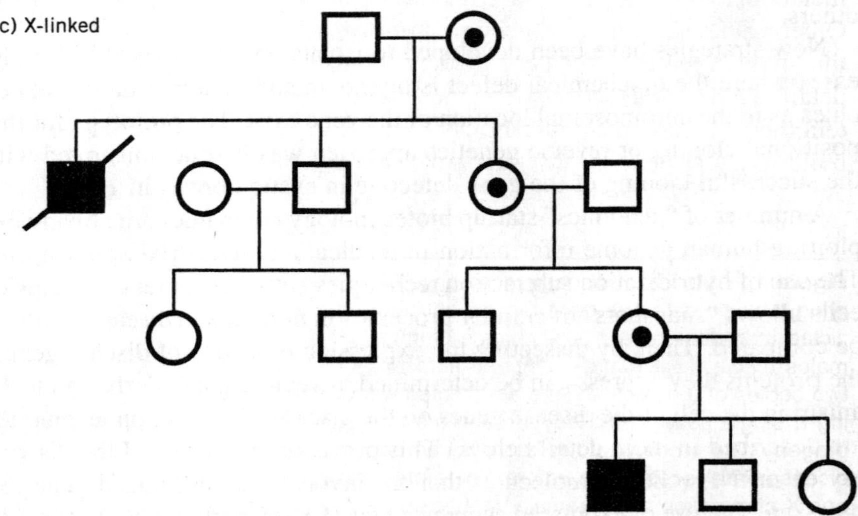

(c) X-linked

Figure 1.2

identifying the mature mRNA transcripts and then reverse translating them into cDNA). The cDNA sequences are inserted into plasmids and used to transform bacteria. Selection of clones for sequencing followed by sequence comparison allows one to determine the abundance and expression of particular genes in the different tissue types. This information is valuable when attempting to determine which genes get turned on during the onset of a genetic disease, or which genes get expressed differently as a result of foreign invasion. Sifting through ESTs will also help determine the distribution of protein-coding genes on chromosomes, generate genetic and physical maps, and identify disease-causing genes (as the example on identifying the STM2 gene on chromosome 1 in Alzheimer's disease shows below). In a recent study identifying new genes and analyzing their expression patterns, over 174,000 ESTs have been generated from 300 cDNA libraries from 37 distinct organs and tissues (Adams et al. 1995). The ESTs were combined with an additional 118,406 ESTs from the dbEST database, yielding over 29,000 tentative human consensus (THC) sequences as well as over 58,000 nonoverlapping ESTs. From the over 70,000 previously unknown human genes identified, those sharing significant similarity with genes known function identified approximately 40% that play a role in basic energy metabolism, cell structure, homeostasis, and cell division, with 22% involved in RNA and protein synthesis and processing and a further 12% with cell signaling and communication. In a similar manner, specific diseases being targeted by commercial concerns include genes that predispose one to breast cancer, melanoma, prostate and colon cancers, heart disease, and hypertension. Other commercial uses of genome information include diagnostics, use of the "wild-type" gene for gene therapy (human gene therapy is described in detail in Chapter 2), developing antisense oligonucleotides for therapy, protein therapy using recombinant DNA, and using genes or gene fragments to build assay systems to discover small molecule mimetics with therapeutic

---

Figure 1.2. Modes of disease inheritance. *Autosomal dominant:* All individuals carrying a single defective gene (X) are afflicted with disease. The children from an afflicted individual have a 50% chance of inheriting the trait. Males (squares) and females (circles) are equally affected. *Autosomal recessive:* Only individuals carrying two copies of a defective gene are afficted with disease (fully shaded). Those carrying a single copy (half-filled) are carriers that can pass the defective copy to their progeny. The progeny of two carriers can become affected. *X-linked:* Most X-linked diseases are recessive in nature. Therefore females may carry a defective gene on one X chromosome and remain unaffected (dotted circle). Once the gene is passed onto a male, it essentially becomes dominant due to the absence of a second X chromosome. Thus males will be afflicted (filled square) or in some cases such as Duchenne's muscular dystrophy will die too young (slashed square) to reproduce.

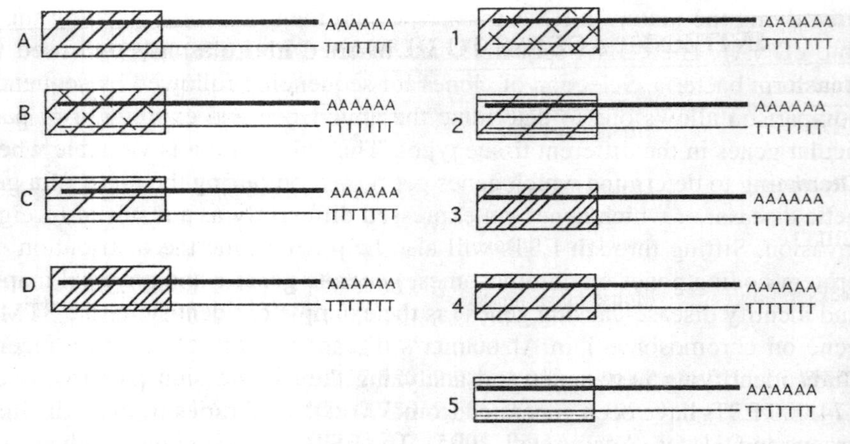

Figure 1.3. Expressed sequence tagging (EST) strategy. For a given sample, mRNA populations (thick lines) are copied to cDNA by reverse transcriptase using a single oligo d(T) primer from the mRNA poly(A) tail. The cDNA inserts of several thousand clones in the library are partially sequenced from one or both extremities (boxes), each sequence (labeled *A–D* in tissue 1 and 1–5 in tissue 2) representing an expressed sequence tag (EST). In the example shown, is a sequence comparison of the ESTs from tissue 1 and from tissue 2 that reveals matches between *A* and 4, *B* and 1, *C* and 2, and *D* and 3, and one EST (5 in the figure) that is unique to tissue 2 (after Ashton et al. 1966).

potential. (These procedures are covered in more detail in the subsequent chapters.)

Clinical genetic research is shifting in emphasis from monogenetic diseases (sometimes paraphrased as "one gene, one polypeptide, one disease") to the study and analysis of polygenic non-Mendelian inheritance common of many diseases. In part this has been achieved by the sharing of genome mapping and positional cloning data. The distribution over the Internet of the Human Gene Map produced by an international collaboration between CEPH, Washington University, the University of Utah, and other sites has been a major step forward.

Clearly the HGI offers an unparalleled opportunity to rationally develop therapeutic molecules against a number of genetic diseases, in addition to those diseases with a genetic component. Although some diseases require the interaction of many genes to reveal their pathology and others are the result of a still more complex interplay between genetic factors, pathogenic factors,

INTERNET ACCESS TO HUMAN GENETIC MAPS

Source	Type	Access
GDB	Human loci data	help@gdb.org; ftp.gdb.org WWW:**http://gdbwww.gdb.org/**
Genethon	Linked CA-repeat markers	FTP:ftp.genethon.fr WWW:**hhtp://www.genethon.fr/**
CHLC	Genotypes, linkages	info-server@chlc.org; ftp.chlc.org WWW:**http://www.chlc.org/**
Jackson Lab.	Mouse mapping data	mgi-help@informatics.jax.org FTP:ftp.informatics.jax.org WWW:**http://www.jax.org/**
CEPH	YAC physical maps	ceph-genethon-map@cephb.fr FTP:ceph-genethon-map.genethon.fr WWW:**http://www.cephb.fr/bio/** ceph-genethon-map.html
Whitehead Inst.	CA-repeats in mouse	genome_database@genome.wi.mit.edu FTP:genome@wi.mit.edu
	Human genome STS map	WWW:http://**wwwgenome.wi.mit.edu**

Source: CHLC, *Science* (1994) 265:2049–2054.

and environmental factors, it is clear that genome-based medicine offers the single best hope in creating safe, effective, targeted therapies. The remainder of this chapter considers the technologies of mapping, sequencing, and bioinformatics and their application to studies of genetic diseases of active interest.

1.2. STRATEGIES FOR MAPPING THE HUMAN GENOME

At the start of the 1980s, an international team of scientists began to use human restriction fragment-length polymorphisms (RFLPs), variations in the size of human DNA fragments generated by standard restriction enzymes, as a means of mapping the human genome. This approach has been applied to families comprising many siblings and an available extended family for which a well-documented family linage exists. The CEPH (Centre d'Etude du Polymorphisme Humain) laboratory in Paris served as the center for DNA RFLP samples from 40 large families from France, Venezuela, and the United States. These data have served as the starting point for an international effort to comprehensively map the human genome at centimorgan (cM) density. (The centimorgan is a measure of genetic linkage corresponding to approximately 100 kilobases of DNA).

Another step in mapping the genome was the discovery of many mini- and microsatellites whose varying number of sequence repetitions allow the definition of many new and informative hereditary polymorphisms. The frequency and relatively homogeneous distribution of microsatellites makes them ideal for human gene mapping, especially since they are easily genotyped using PCR amplification. The development of about 4000 RFLPs and 2000 microsatellites has resulted in a human gene map with mean density of 0.7 cM. This map is allowing researchers to construct a physical map of regions of the genome with cosmid-sized overlapping DNA contigs cloned into yeast artificial chromosomes (YACs).

The expansion of computer-based methods in the 1980s allowed the evaluation of multiple linkage markers as well as in the number and type of DNA-based markers. The use of micro- and minisatellites as markers (sometimes called variable number of tandem repeats, or VNTRs) provided a source of markers of high information content that more or less saturated the entire genome. The first genomewide linkage maps emerged in the late 1980s were based primarily on the 61 families whose linage was maintained at CEPH in Paris. (The CEPH database enabled markers from around the world to be pooled and studied on a common set of families). These maps and markers are now routinely used as the initial step for positional cloning of disease genes, such as the isolation of the gene for Huntington's disease described later in this chapter.

Currently comprehensive genomewide human linkage and marker maps are being developed by three main groups, together with the contribution of several other investigators. A group at Genethon in Paris is focused on maps based on markers that contain the CA repeat motif, while the Cooperative Human Linkage Center (CHLC) in the United States is constructing maps based on di-, tri-, and tetranucleotide repeats. Chromosome-specific research is being conducted by the NIH/CEPH consortium and by EUROGEM.

Development and Applications of Human Genetic and Physical Maps

A physical map of the human genome at 100-kbase spacing requires about 30,000 ordered sites distributed evenly among the genome. The most popular method for physical mapping currently is sequence-tagged site (STS) mapping, resulting in many overlapping partially completed maps (Hudson et al. 1995). Here genetic positions are mapped by identifying unique sequence fragments that unambiguously pinpoint a given gene or genetic region. The use of PCR to amplify any chosen STS site is essential to this procedure of physical map creation. STS maps then, are collections of labeled, specific primer pairs used to provide unique reference points on genetic and physical maps. Mapping is then reduced to determining the order of the STSs, inferred from the data on the STS content in random segments of the human genome present in either YAC library clones or in panels of human-rodent hybrid cell lines (sometimes termed *radiation-hybrid* or RH lines). Currently the average spacing between STSs for the entire genome is about 200 kbases. (Small fractions of the genome have been mapped at a much finer resolution.) The primers may be used as templates for PCR amplification by any laboratory. Direct transfer of the reagents needed for mapping is thus simply the passing of information containing the primer sequences rather than any biological materials. A combined genetic and physical map utilizing both genetic and physical markers has been published (Cohen et al. 1993), and a STS-based map at a resolution of 200 kbases has recently appeared (Hudson et al. 1995).

The existence of genetic and physical maps now makes genomewide searches for Mendelian disorders possible. There are currently 3617 STRP (short tandem repeat polymorphism) class markers, corresponding to one marker every 10^6 bp throughout the genome. The correlation of linkage to genetic markers is determined by the lod score or logarithm-of-odds ratio, which is a measure of the linkage of a given set of data. Once an initial localization is identified by means of such genomewide maps, subsequent steps may be taken to narrow the localization and eventually select a specific interval where DNA cloning can be carried out. Searches for particular genes may involve characterization of conserved sequences or CpG islands, studies of tissue-specific expression (e.g., ESTs), direct searches for cloned sequences by hybridization (differential display; see Fig. 1.4), or direct large-scale sequencing. Note that the correlation between genetic and physical distance has not yet been explicitly defined throughout the genome. Variation in recombination rates per unit of physical distance vary by a factor of 100 and distort the perceived distance a given point may lie from a locus or marker of interest. Nonetheless, the recent successes in mapping the susceptibility genes for diabetes and identifying the gene locus for two forms of breast cancer confirm the utility of dense linkage maps. In addition physical methods still need to be developed that are capable of high-level ordering of cosmid-sized fragments (up to 10^6 bp).

A physical map of the human genome has been reported consisting of 15,086 STSs, at an average spacing of nearly 200 kbases (Hudson et al. 1995). This corresponds to 6193 loci, providing radiation-hybrid coverage of 99% and physical coverage of 94% of the human genome. The map is expected to provide a scaffold for large-scale sequencing of the entire human genome.

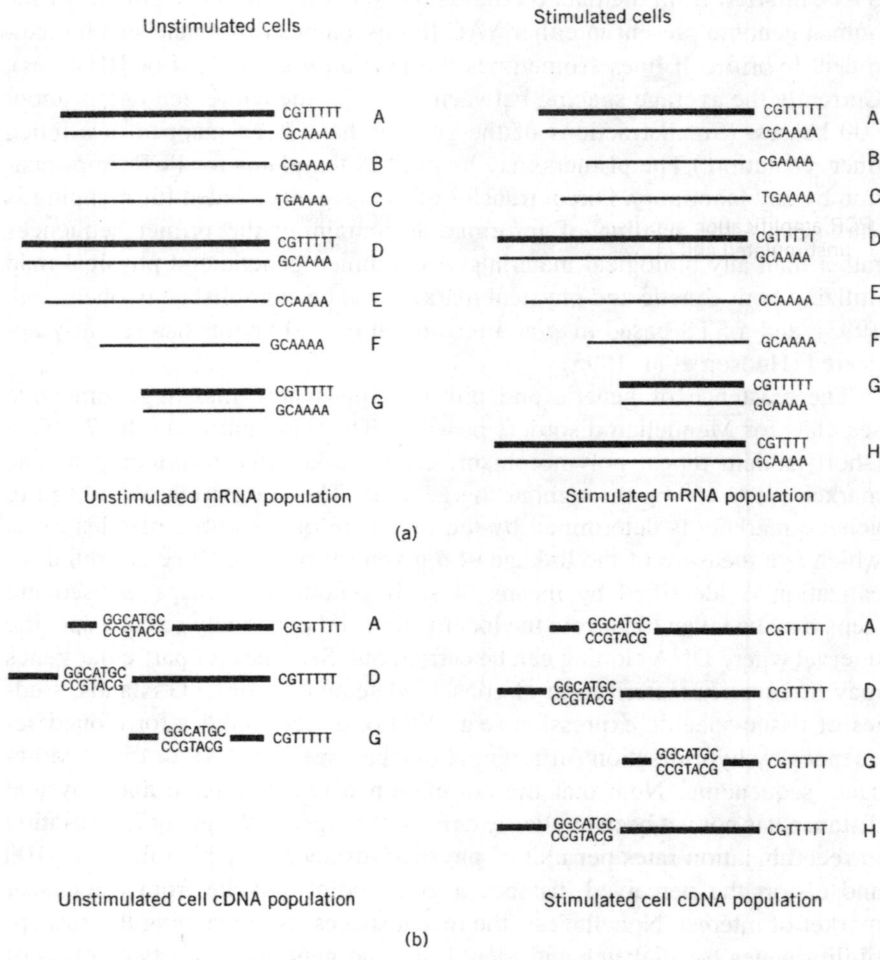

Figure 1.4. Differential display. The differential display technique distinguishes genes in normal and diseased tissues by spatial immobilization or gel electrophoresis. Shown is the method of identification by gel electrophoresis (a) *Reverse transcription:* For a sample of stimulated (or diseased) and normal cells, mRNA subpopulations (thin lines) are copied into cDNA (heavy lines) using a downstream primer of the form 5′-poly(T)-XX-3′, where XX is one of deoxycytidilate, deoxyadenylate, or deoxyguanylate. (b) *Upstream primer annealing:* Each sample is annealed to a set of arbitrary upstream primers.

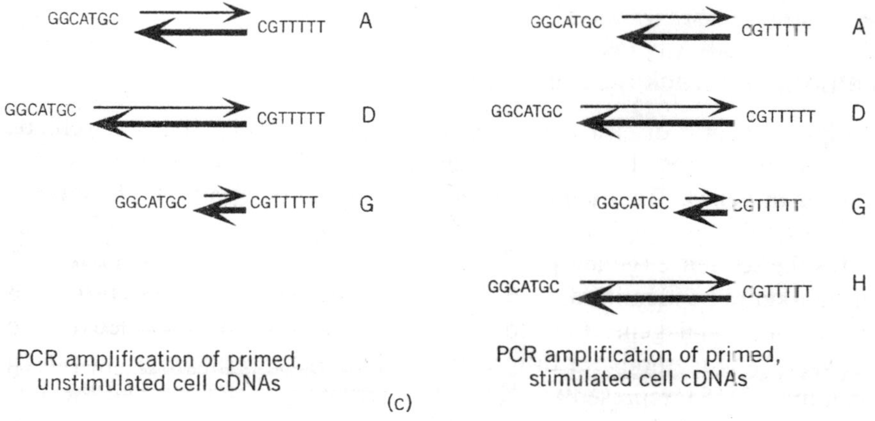

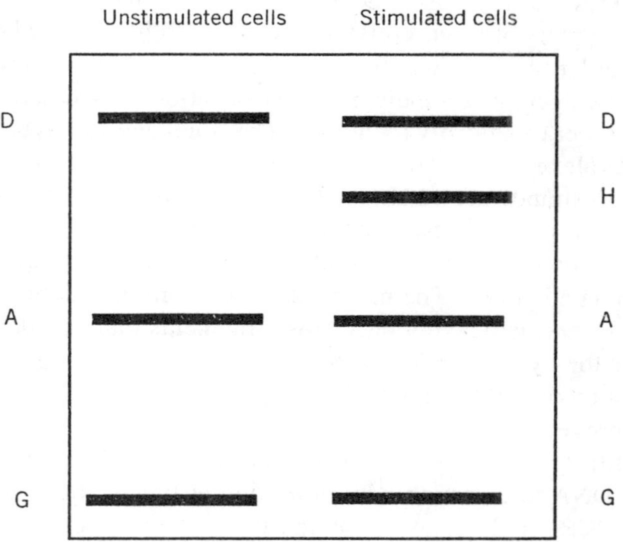

Polyacrylamide gelelectrophoresis of PCR products

(d)

Figure 1.4 (Continued). (c) *PCR amplification:* PCR is performed on each of the cDNA products from each sample. In the example shown, both mRNA populations give rise to products A, D, and G, while product H is generated only in the stimulated cell population. (d) *Gel electrophoresi:* The radio-labeled PCR products are visualized by polyacrylamide gel electrophoresis, revealing product H as unique to the stimulated cell population (after Ashton 1966). Variations of this technique are known as genome midmatch scanning (GMS), representational difference analysis (RDA), or transcript imaging (TI).

Methods for Tracking Disease Genes

Complex genetic diseases such as diabetes, severe obesity, and schizophrenia are the result of the individually relatively weak influences of a number of genes. Traditional linkage analysis (associating the phenotypic effect of a disease gene with a marker that is always inherited with it) requires the sequence (genotype) of each marker for each individual carrying this marker to be identified, one marker at a time. The weaker the influence of the individual gene, the more families that must be included in the analysis. New methods for differential display such as genomic mismatch scanning (GMS), representational difference analysis (RDA), and the serial analysis of gene expression (SAGE) allow the entire genome to be scanned for variations. Differential display is illustrated schematically in Fig. 1.4 (see pages 16–17). GMS detects similarities between pairs of genomes, allowing pairs of disease-affected relatives to be analyzed for an entire family in order to identify regions that are identical among affected individuals. The two genomes are digested by restriction enzymes, and one is methylated in order to uniquely identify it. The fragments are hybridized and only heteroduplexes (those that contain hybrids of one methylated and one unmethylated strand) are retained. Mismatches between the hybrid fragments indicative of single base mutations within the targeted disease are recognized using the *E. coli* mismatch repair system to “nick” the “newer” unmethylated strand. The nicked fragments are then digested using another DNA enzyme, leaving only those fragments that are identical. By immobilizing the hybrid DNA fragments onto an ordered array, the efficiency of detection of these fragments in the pairwise comparisons can be significantly improved.

In RDA the differences between two genomes are identified. After subjecting genomic DNA to digestion with restriction enzymes, the fragments are subjected to PCR amplification, resulting in *amplicons*—smaller DNA fragments sampled from the original genome that are easily hybridized to another genome. By having the second genome present in excess to the first, *in vitro* hybridization guarantees that essentially all of the amplicons from the first genome will hybridize to fragments in the second and not to themselves. Although the amplicons typically represent only about 10% of the genome, using different restriction enzymes in the initial genome digest results in different amplicons and hence a different sampling of the genome. Once again the results of the hybridization are represented on immobilized ordered arrays allowing the location of the amplicon that contains mutations in the DNA sequence that result in a particular disease. The process is being used to study mutations involved in cancer, by comparing DNA from tumor cells to the DNA in healthy cells from the same patient.

SAGE relies on the fact that small sequences (as few as 9 bp) can be used to identify the vast majority of genes provided that the sequence is selected from the same place in genes compared between healthy and diseased tissues. Active mRNAs from both healthy and diseased tissues are first extracted and reverse transcribed into cDNAs that are tagged with biotin at their 3′- ends. The cDNAs are then digested, and the fragments isolated on biotin-streptavidin beads. A second restriction enzyme is then used to cut out fragments containing a chosen 9-bp sequence. These tags are then spliced together, PCR amplified, and sequenced from both the healthy and the diseased tissue. Comparison of these genetic *bar codes* allows immediate identification of those genes expressed only in one tissue but not the other (Kinzler et al. 1995).

Medicinal Applications of a Linkage Map of the Mouse

High-resolution genetic linkage maps of the mouse genome at a resolution of about 2 Mbase have been generated by a collaborative project of scientists across the United States. The mouse genome is of particular interest because of the use of mice as models to mimic human diseases and identify strategies for interventional therapy. (The use of mice to study immune dysfunction is described in detail in Chapter 5.) Although mice and humans diverged more than 75 million years ago, they are remarkably similar genetically. Their early embryological development also shows striking similarities. Inbred strains of mice serve as important models for human genetic diseases such as anemias, autoimmunity, and immune dysfunction (the SCID mouse), neurological disorders, birth defects, cancers (the MIN mouse), diabetes (the NOD mouse), and others. Recently the development of transgenic and embryonic stem cell technology has made it possible to express virtually any gene in any mouse tissue and create targeted germ line gain-of-function or loss-of-function mutations (see Chapter 2). The ability to introduce yeast artificial chromosomes carrying several hundred kilobases of genomic DNA into mouse germ line cells now provides us with unparalleled opportunities for genome analysis of new mouse models for human disease.

Of the 2616 mouse loci that had been mapped by 1994, 917 have homologies identifiable in humans. These loci mark 101 segments of conserved linkage homology. About 61% of the mouse genome is accounted for in this comparative map between human and mouse. Several examples are found to exist of linkage conservation across a human centromere, such as human chromosome 20 on mouse chromosome 2 and human chromosome 17 on mouse chromosome 11. The pattern of variable and conserved segments indicates that multiple chromosome rearrangements must have occurred since the divergence of the linage leading to human and mice. Approximately 150 rearrangements are thought to have occurred since this divergence. Rearrangements may be through translocations, inversions, insertions, and other complex rearrangements. In some cases they have even lead to changes in gene order in conserved segments. One application of comparative mapping is the transfer of linkage from map-rich species such as humans and mice to map-poor species such as cow, pig, and sheep. Another application is the analysis of complex traits, whereby identification of genes responsible for a trait may be easier in mice than in humans. Once a candidate disease gene is identified, it can be used to screen homologous regions in humans to see if they are linked to the corresponding human genetic disease. Certain mouse strains show striking variation in their susceptibility to diabetes, epilepsy, cancer, bacterial and viral infections, and obesity. This feature makes them attractive candidates for the genetic dissection of these complex multigenic diseases. There is also an inherited variation in physiological parameters such as skeletal morphology, blood pH, response to drugs and hormones, immunological responses, and life-span. Genetic dissection of such polygenic traits requires following the simultaneous inheritance of markers spanning the entire genome to identify regions that together account for the phenotype. A comprehensive genetic dissection of the factors causing type I diabetes in the non-obese diabetic (NOD) mouse has begun, and at least 11 putative diabetes genes have been identified. Similarly genetic factors have been dissected for epilepsy and familial colon cancer. It is hoped that these studies in mice will allow the dissection of multigene diseases in humans without the need to isolate as rich or complex a set of markers.

1.3. THE ROLE OF BIOINFORMATICS IN GENOME-BASED THERAPY

The role of bioinformatics in molecular biotechnology is to provide an informational framework that integrates the disparate data obtained about the workings of the living cell by methods such as mapping and sequencing. This information can then be used as a starting point for the development of new or improved molecules of therapeutic or clinical importance. Armed with a workstation, suitable software and databases, and access to the Internet, a biomedical researcher can begin to answer a host of important questions such as: Have I isolated the disease gene I wanted? What function does (might) it have? What other genes is it related to? How much of it do I need to sequence? How many proteins does it code for, and what function might they have? What can I predict about the structure and action of these proteins? What regulatory elements might exist in the gene? What interventional strategies can I propose based on this information? Indeed the modern genomic researcher can sometimes answer questions about a particular disease state even before any data have emanated from their own laboratory. Analysis of the volumes of gene sequence, physical map, linkage (genetic) map, and other cellular and clinical data now filling the different databanks daily gives the researcher the ability to determine how much of a particular system has already been determined at the molecular level and how representative the data is of the entire system (Waldrop 1995). This can direct the scientist to those aspects of the system that remain to be completed, or even to study a different system instead. Conversely, this information can also tell scientists how likely they are to identify the function of any novel genes they may isolate and sequence based on data that already exist about that particular system.

Formally, bioinformatics encompass the use of software, databases, and networks for gene and open-reading frame (ORF) identification; database homology and pattern searching with both DNA and protein sequences; comparative sequence analysis and multiple sequence alignment; protein structure prediction and the mapping of functional sites; protein homology modeling and "inverse" folding as a means of probing protein structure and function; and the discovery or design of drugs against genes or their products. For example, comparative sequence analysis of polymorphisms or other variations in a family of genes can pinpoint specific regions against which to target drugs. The ability to express gene products as soluble receptors allows *in vitro* screening methods to be constructed that can be used to develop drugs against the protein products of these genes. These methods require access to a comprehensive library of gene and protein sequences. The principal databases of the HGI are GenBank and Genome Sequence Data Base (GSDB) for gene sequences, PIR for protein sequences, the Genome Data Base for chromosome information, the sequence-tagged site database dbSTS, and the expressed sequence tag database dbEST. The latter is a database of public complementary DNA (cDNA) sequences that has recently reached over 50,000 in number, corresponding to nearly 85% of human disease sequences positionally cloned to date (Boguski et al. 1994). In Europe these databases exchange data with the EMBL database of gene sequences, the SWISS-PROT database of protein sequences, and the CEPH database of polymorphic markers identified in human families. Each of these databases is accessible and searchable via the Internet. Japan archives DNA sequences obtained on the Asian subcontinent in the DNA DataBank of Japan (DDBJ). GenBank, EMBL, and DDBJ exchange sequence data among each other nightly. Another database of commercial importance is the GENESEQ™ database of patented DNA and protein sequences published by Derwent, Inc./IntelliGenetics.

Historically the first organism to have its genome completely sequenced was the bacteriophage phi-X-174 (Sanger et al. 1977). Since then, other organisms whose genomes have been completely sequenced include that of cytomegalovirus (Bankier et al. 1991), vaccinia (Goebel et al. 1990), and variola (Massung et al. 1993), as well as the mitochondrial and chloroplast genomes of *M. polymorpha* (Oda et al. 1992). Active genome-sequencing projects exist for the fruit fly *Drosophila* (Hartl and Palazzolo 1993), the bacteria *E. coli* (Sofia et al. 1994) and *B. subtillus* (Glaser et al. 1993), the yeast *S. cerevisae* (Levy, 1994), the nematode worm *C. elegans* (Sulston et al. 1992), and of course *homo sapiens* (Bodmer et al. 1994). These projects were all sequenced primarily by the sequencing of clones derived from well-mapped restriction fragments, or from lambda or cosmid clone libraries. The largest single clone that can be readily sequenced by these methods is still about 40 kbases. The primary limitation on sequencing larger clones is the lack of robust computational approaches that could efficiently assemble tens of thousands of independent, random (shotgun) sequences into a single assembly. Recently speculation has arisen that entire genomes may be sequenced by shotgun approaches, thanks in large part to new developments in bioinformatics software and hardware (Fraser et al. 1995). In particular, computer programs have been designed that can handle the accurate recognition and correct alignment of the ends of overlapping fragments in order to "knit" together the entire genome sequence. The degree of coverage needed to sequence an entire (contiguous) genome by the shotgun approach may be determined a priori, and an estimate of the number of gaps for a particular coverage rate that will need to be filled identified (Lander and Waterman 1988). Clearly the accuracy of the sequence generated by this approach must be carefully verified by comparing the number of sequence errors, ambiguities, and possible frameshifts in those regions where sufficient overlaps exist. In particular, errors or ambiguities due to sequencing compressions at the ends of gels can lead to the propagation of ambiguities in the overlapping fragments, quickly generating nonsense consensus sequences when many contigs are overlapped. Bioinformatics software for large-scale shotgun sequencing must be able to verify the random generation of shotgun libraries; handle at least a sixfold coverage; assemble all sequence fragments (contigs) and identify repeat regions; order all the contigs and provide templates for gap closure; fill-in gaps with sequencing data obtained from primer walking; allow editing and visual inspection of the sequence and manually or automatically resolve sequencing ambiguities including frameshifts; and identify and describe all (predicted) coding regions, start/stop codons, control regions, operons, and regulatory regions.

The HGP is currently generating about 250 to 500 kbases per day and this is expected to increase to 10 Mbases per day within the next five years, assuming only incremental improvements in sequencing technology. The need to analyze this much data in real time requires standardized databases to collect, manage, and distribute the data and a collection of networks connecting each of the data centers both nationally and internationally. Genetic (nonsequence) databases are currently in an international collection classified by organism, such as Flybase *(Drosophila)* in the United Kingdom, the Genome Data Bank (humans) and GBASE (mouse) in the United States. A Genome Sequence Data Base (GSDB) has also been set up specifically for whole genome sequences. Databases have been designed to be interoperable (available over heterogeneous networks such as the Internet) and decentralized to permit interactive searching from remote locations. Computer systems and software are now required for virtually all aspects of genome studies, from data collection and analysis to data management and distribution. The analysis of protein expression in cells as a function of gene expression has led to new tools such as laser desorption and electrospray mass spectrometery (MS) to measure and identify proteins on 2D gels, coupled with expert system software to unambiguously and automatically identify the expressed proteins. Because the Genome project is generating information of a type not typically found in biology, novel electronic publishing systems, including bulletin boards (the BIOSCI™ newsgroups) and hypertext links are being developed. The use of sophisticated computer technologies such as the subset of the Internet known as the World Wide Web (WWW), and software such as Mosaic™ and NetScape™ will be instrumental in disseminating this information. Collaborative and community databases are also being formed, although some of the larger ones suffer from political turf wars. A clear need for on-line documentation and user support has emerged, or else bioinformatics risks being balkanized into segments where the best data are available only to the large genome centers that can afford full-time bioinformatics staff. In the United States the National Center for Biotechnology Information (NCBI), a division within the National Library of Medicine (NLM), has been established by an act of Congress to oversee all aspects of bioinformatics for the HGI. In Europe the EMBL has established a European Bioinformatics Center in Cambridge, England, to oversee all bioinformatics needs for the European community. To complicate matters, a number of commercial concerns are also sequencing portions (typically cDNA fragments) of the human genome for commercial gain and placing their data in proprietary databases that may or may not be accessible to the public. A vigorous scientific and economic debate is currently in progress regarding the value of maintaining proprietary sequence information (e.g., to allow the patenting of genes or gene fragments of possibly unknown function) versus releasing these data to the public domain for access by all. (The recent acquisition of the rights to commercialization of therapeutic agents developed from the discovery of the leptin gene, conferring a predisposition to obesity, by the Amgen Corporation for U.S. $20 million illustrates the competitive value placed on this information by biopharmaceutical companies.)

Future developments in bioinformatics might include linked entries from different databases such as that demonstrated by the experimental network of databases known as ACEDB (AC. *Elegans* DataBase); new tools for the discovery of information from the data in current databases such as the recognition of promoter sequences or gene start/stop sites; rapid and straightforward sharing of bioinformatics data; and the ability to relate DNA information to the complex control networks that regulate the immune system, intracellular signaling, and cell development. Higher-order databases of important value might include databases of enzyme activities, taxonomy databases, plant genetic databases, quantitative structure activity relationships (QSARs), metabolic pathways, and drug interaction sites. The uses of databases and libraries pertinent to drug discovery and rational drug design are discussed in Chapters 3 and 4.

1.4. ANALYSIS OF HUMAN DISEASE GENES

Some examples of common diseases being tackled through the use of genome-based medicine are described below. The examples are intended to illustrate some of the power and diversity to which this technology can be applied.

Huntington's DIsease

Huntington's disease (HD) is a progressive physical and mental deterioration attributed to a single gene defect that manifests itself to those in the mid 30s to mid 40s, when individuals typically already have had children. HD is carried as a dominant disorder in a simple Mendelian fashion, so victims of the disease have a 50% probability of passing it to their offspring. By analyzing the inheritance of genetic markers in families afflicted with the neurological disorder, scientists have finally isolated the gene. The search, over a ten-year period, illustrates some of the issues of hunting down disease genes that the HGI hopes to obviate with access to complete gene maps and gene sequences. An international consortium of scientists from the United States, England, and Wales together identified the gene to a position at the tip of chromosome 4, flanked to within 250 kbases by markers. The gene defect is yet another illustration of the triplet repetitive sequence that has been found to be the source of the genetic diseases fragile X syndrome and myotonic dystrophy. Analysis of markers from HD patients suggested a 2.2-Mbase sequence near the telomere of chromosome 4 and a particularly strong correlation to a 500-kbase segment within this region. Cloning and sequencing of a transcript labeled IT15 showed a 210-kbase open-reading frame, coding for an unknown protein that also had numerous CAG triplet repeats at its 5′ end, that is, the unstable portion of the gene responsible for causing HD. Comparison of this ORF from 150 HD patients with the same gene is unaffected patients showed that the former have between 42 to 86 of the CAG repeats, with the length of the repeat sequences correlating with the onset and severity of the disease.

Future research hopes to determine the function of the protein coded by the gene and how it is affected by the triplet-repeat mutations. The gene is known to be active throughout the body, yet the mutant gene product affects the cells of the brain far more significantly than elsewhere. By genetically engineering mice with the HD gene and the extra CAG triplets, a series of experiments can be performed to determine the effect of extra triplet repeats on neuronal or other cell functions (see Sec. 2.4)

Identification of the gene does present a fairly simple and conclusive diagnostic test, however, since individuals who have a parent with HD stand a 50% chance of inheriting the gene and also coming down with the disease. Current testing involved comparing markers from many family members to determine if a person suspected of having HD inherited the markers that are closely linked to the gene. Here there are significant psychological and personal risks involved, including the effect of the disease on the person's job, insurance, or relationships. A further complication arises in that the extent of the triplet repeats correlates with the degree and onset of the disease manifestation. In the United Kingdom the diagnostic test is currently only given in the presence of physicians trained to counsel patients on the effects of the disease and how it will affect their quality of life.

Cystic Fibrosis

Cystic fibrosis (CF) is a recessive genetic disease with a carrier frequency of about 4% in Caucasians (Boat et al. 1989). The disease is inherited in an autosomal recessive fashion; that is, those heterozygous for the disease carry one normal allele and one mutant allele and are asymptomatic carriers, whereas the child of two carriers has a one in four chance of being affected by CF. The pathology of the disease manifests itself as an imbalance in salts in the cells of the epithelial membranes of the lungs, resulting in a buildup of cells and sputum that clogs these passageways. Cellular damage can extend to the pancreas, kidneys, and vas derferens. The discovery that the normal efflux of chloride ions across epithelial cell membranes of the respiratory glands in response to elevated cAMP levels is lacking in cells from patients with CF was instrumental in the discovery and isolation of the CF gene (Sato and Sato 1984). Positional cloning was used to identify the CF gene by first mapping it to chromosome 7 through the use of linkage analysis of polymorphic DNA markers in multiple affected individuals. Subsequent refinement placed the gene in an ∼1.5 Mbase DNA interval. Through the use of chromosome jumping and walking (Rommens et al. 1989), a collaborative effort identified a candidate transcript for the gene that was expressed in the sweat glands, lungs, and the pancreas (Riordan et al. 1989). The gene was identified as 250 kbase pairs long, corresponding to an mRNA transcript of 6.5 kbase pairs, and a protein of 1480 amino acids (Kerem et al. 1989). Proof that this transcript was the gene came from identification of a mutation accounting for nearly 70% of mutant alleles in CF patients. This mutation, a 3 bp deletion in exon 10, results in the loss of a single amino acid, phenylalanine 508, designated F508 (Kerem et al. 1989). A large number of other mutations (over 170) have also been discovered, occurring in frequencies varying with the geographic and ethnic subgroups in which they have been found. The ability to detect the most common 80–90% of the mutations that characterize CF has made it possible to launch general population screening programs to identify couples who may be carriers (Caskey et al. 1990). In addition gene therapy trials have begun in both the United States and the United Kingdom, using different vectors to attempt to cure this condition (Rosenfeld et al. 1990; Cotten et al. 1990). These studies are described in more detail in the following chapter.

The gene product has been named the cystic fibrosis transmembrane conductance regulator (CFTR). Its function was first identified by homology searches of databanks of gene and protein sequences—a triumph of the emerging science of bioinformatics. The gene showed a striking similarity to a family of proteins involved in active transport across cell membranes (as detected by computer-based similarity searches of sequence databases) referred to variously as the ATP binding cassette (ABC) family or the TM6-NBF family (Hyde et al. 1990; Riordan et al. 1990). CFTR has been shown to conduct chloride ions, and evidence exists that other ions can traverse the CFTR pore without ion–ion repulsion taking place (Anderson et al. 1991; Barasch et al. 1991). The primary expression of the gene is the exocrine gland, where mutations that disrupt the CFTR protein, convert it from a multi-ion pore into a single-ion pore with reduced conductance in patients with the disease. Gene transfer of the wild-type CFTR coding region has been demonstrated to be sufficient for correcting the chloride channel defect. Mutations to residue 347, an Arginine in the pore wall, have been identified with reduced pore conductance, particularly of chloride ions, while the mutation to Phenylalanine 508 is now thought to cause the channel pore to remain closed for a greater portion of the time than in the wild-type protein, disrupting the chloride efflux. This conductance is essential for fluid secretion and salt absorption in the wet epithelia of the lungs that are affected by the disease. When secretion falls below a minimum, blocked exocrine glands can result. A cascade of effects that are still poorly understood results in significant damage and even destruction of the pancreas and vas deferens and makes the lungs susceptible to repeated infections.

The identification of the gene and its product in CF lead to new possibilities for improved drug therapy. Interventional strategies can now be directed at different stages of the disease pathology. The abnormal CF gene can be corrected by providing a normal gene via gene therapy, and the mutant protein can be compensated in the same way by expression of the transgene in epitheleal cells. A potassium-sparing diuretic drug named Amiloride is in trials to block the increased NA^+ uptake and decreased Cl^- efflux that is characteristic of modified CFTR channels exhibiting abnormal salt transport. A major clinical problem of CF patients is the large volume of mucus produced in the lungs that is heavily infected with bacterial organisms and contains many white blood cells, many of which die and release their DNA, further increasing the viscosity of the mucus. Digestion of the DNA component of the mucus by deoxyribonuclease (DNase), a DNA-cutting enzyme, has been shown to result in better clearance of mucus from the lungs. Genentech, Inc. has now successfully cloned and sequenced recombinant human DNase I and shown its efficacy in clinical trials, and DNase I is now commercially available as Pulmozyme™, provided as an aerosol spray to CF patients. The development of a gene therapy product for CF and the clinical progress made so far are discussed in detail in the following chapter.

Retinosis Pigmentosa

Retinosis pigmentosa (RP) describes a genetically and clinically heterogeneous group of human inherited retinopathies that destroy the rod and cone photoreceptor neurons, leading to loss of sight and eventually night blindness in about 1 in 3000 adults. RP is found to be inherited in three distinct fashions: autosomal dominantly (adRP), autosomal recessive (arRP), and in an X-linked fashion (x1RP). arRP seems to be the most frequent, representing about 50% of families afflicted with the disease. By contrast, adRP and x1RP are together present in about 10–15% of RP families.

The rapid emergence of a genetic map of the human genome has facilitated the localization of some of the genes for RP. The use of PCR with highly polymorphic genetic markers based on sequence repeats has sped the localization of these genes—together with methods for rapid identification of sequence variations such as single-strand polymorphism electrophoresis

(SSPE), heteroduplex analysis, denaturing gradient electrophoresis—and direct sequencing of amplified DNA products have allowed the localization of 2x1RP genes to the short arm of the X chromosome. Mapping of 3 adRP genes to the long arm of chromosome 3 and the pericentric region of chromosome 8, and mutations in the rhodopsin gene mapped to 3q and the peripherin RDS gene mapped to 6p, have been implicated as causative in some forms of adRP. Both of these genes code for proteins that traverse the outer disc of the rod photoreceptor cells.

The first genetic linkage studies of RP were established when a polymorphic DNA marker named L1.28 was associated with RP in five families with the X-linked form of the disease. The position of L1.28 on the X chromosome had previously been determined by fluorescence *in situ* hybridization (FISH), whereby a fluorescent DNA marker is allowed to hybridize to its homologous sequence on the intact chromosome and the fluorescence detected by microscopy. The RP locus mapped to the short arm of the X chromosome close to the centromere. Several patients with deletions to the short arm proved valuable in further pinpointing the locus of the disease gene. The positions of the deletions enabled the precise location of the RP locus to be determined. The RP2 and RP3 genes have now been localized by a combination of linkage and mutagenesis studies to specific portions of the X chromosome.

Studies of mutations in families suffering night vision problems (nyctalopia) led eventually to a gene located on the long arm of chromosome close to DNA marker C17 (D3S47). This marker had an exceptionally high lod score (20) with the known position of the rhodopsin gene. Searches for mutations in the rhodopsin gene in patients with adRP led to discoveries in the early 1990s of a Proline 23 to Histidine 23 substitution in up to 15% of all adRP patients in the United States. This mutation is absent in European populations, suggesting a founder effect in the U.S. population. Well over 30 additional mutations within the rhodopsin gene have now been identified in cases with adRP. Most are point mutations in conserved regions of the molecule, while a few are small deletions or frame shifts resulting in radical changes to the carboxy-terminus of the protein.

Because the rod photoreceptor cells are extremely well studied, prior knowledge of the structure and function of these important cells can be used to direct future linkage and candidate gene studies. In particular, the genes encoding for transducin, rhodopsin kinase, arrestin, and interstitial retinol binding protein (all photoreceptor cell proteins) are just a few of many potential candidates. It is expected that the majority of these candidates will be cloned and sequenced within the next few years. However, little is yet known about how mutations to these proteins result in photoreceptor cell death.

Harnessing expression systems (e.g., the baculovirus expression system for mammalian proteins) will allow large quantities of normal and mutant proteins to be produced and their functions investigated. An animal model, such as the rd mouse, will also allow immediate identification of genes for yet unknown proteins that may also be causative agents for RP. Finally, targeting potential RP genes by homologous recombination (section 2.4) will allow investigation of the cause of the retinal degradation in the chosen animal model.

Alzheimer's Disease

Alzheimer's disease (AD) is a heterogeneous disorder characterized by disruption of selective neuronal and corticoneuronal connections. The pathological characterizations of the disease are neurofibrillary tangles and senile plaques; the main constituents of which are tau protein and amyloid-beta (A-beta) protein, respectively. Abnormal processing of these proteins by means not as yet determined lead to their insoluble forms being deposited in neuronal cells in people afflicted by this disease. Recent evidence points to protein glycation as involved in the development of Alzheimer-type pathology. In some cases of the disease, a heritable component has been identified, a mutation in chromosome 21 in the gene for an encoded membrane glycoprotein known as amyloid precursor protein (APP). Other mutations have been identified in S182 a single gene on chromosome 14 at the AD3 locus and the E4 allele of the ApoE gene (see below).

A single nucleotide change has been identified in APP in those patients that suffer from familial AD. Sequencing of the APP gene from many patients identifies a few similar mutations. The APP protein is a precursor to amyloid beta-protein (ABP). Two features are characteristic of AD: the amyloid fibril (composed from ABP) and the neuritic dystrophy (characterized by tau protein). The ABP fragment has a mass of ~4000 daltons and a length of 42 or 43 amino acids. The protein forms 75–100Å long fibrils which are essentially in the cross-beta pleated conformation, as identified by X-ray diffraction. The mechanism by which an amyloid fragment undermines the viability of the neuron is not well understood, however. One hypothesis suggests that the peptides are toxic to the membranes of the neuronal cells.

Recent studies in 234 people from 42 families with late-onset Alzheimer's have implicated one of the Apolipoprotein genes (ApoE). Almost all those who had two copies of the ApoE4 gene was found to develop the disease. Their overall risk was almost eight times greater than that of people with no copies of the gene. ApoE4 seems only associated with late-onset Alzheimer's, the most common form of the disease (afflicting about 75% of all Alzheimer's patients over 65), whereas previous genes implicated in the disease (beta-amyloid gene on chromosome 21) and another unisolated gene on chromosome 14 are linked to the rare early-onset version of the disease that begins in people under 60. Confirming the genetic findings are the presence of antibodies to ApoE4, which are found in the three lesions characteristic of the disease: extracellular senile plaques, vascular amyloid deposits, and neurofibrillomary tangles found within brain cells. Epidemiological studies point out that the three alleles for ApoE (E2, E3, and E4) are found in different proportions in patients with the disease (where the E4 from predominates) and others (where E3 is the dominant isotype). "Sporadic" Alzheimer's patients (those with no family history of the disease) also show a predominance of the E4 allele (61% vs. 31% in control groups). Further evidence for the role of E4 comes from genetic studies of families with late-onset Alzheimer's, where only 20% develop the disease with no copies of E4; 45% develop the disease if they have one copy of the E4 allele; and nearly 90% develop the disease if they harbor two copies of the E4 allele. Furthermore the average age at onset declines as the gene dose increases, from 84 to 75 to 68, as the dose goes from 0 to 1 to 2. ApoE4 is a well-studied protein known to carry cholesterol in the blood, and these data immediately raise the question of the role of the protein in the disease pathology. ApoE4 is also known to bind beta-amyloid much more strongly than the more common ApoE3 isoform, leading some to speculate that the bound form somehow kills brain cells. In this model ApoE4 acts as a "pathological chaperone" binding the soluble form of the beta-amyloid peptide and making it insoluble and thus more likely to form plaques and lesions that are the hallmarks of the disease. Soluble beta-amyloid has also been shown to bind to Apolipoprotein J, suggesting a delicate balance between two carrier proteins that is disrupted when E4 is the major isoform present. An alternate model suggests that ApoE4 instead interacts with tau protein, the molecule implicated in neurofibrillary tangles where beta-amyloid is not present.

Recently a gene (named STM2—for seven transmembrane gene 2) on chromosome 1 has also been identified as responsible for AD in the so-called Volga German kindreds (Levy-Lahad et al. 1995). The sequence of this gene is found to be highly similar to that of gene S182, identified on chromosome 14, which has also been implicated in familial AD. It was identified in part from a comparison of the published sequence of S182 against a database of expressed sequence tags, which identified another highly similar sequence fragment in the database. A single point mutation in the identified sequence of STM2, the replacement of an asparagine (N) with an isoleucine (I) at position 141 in the sequence (labeled N141I), was identified in every member of the Volga German families exhibiting AD. The asparagine is conserved in both human and mouse S182 homologs, suggesting it plays an important role in the function of the protein.

1.5. THERAPEUTIC GENOMICS IN THE PRIVATE SECTOR

The Institute for Genomic Research (TIGR) is a non-profit privately funded organization that is performing cDNA sequencing of fragments (typically about 400 bp) of the human genome. They claim to have sequenced about 100,000 genetic fragments, corresponding to about half the genes in the human genome. The institute proposes that gene sequences with potential therapeutic application be patented and commercialized by Human Genome Sciences, Inc., a company that has the first right of refusal to all gene sequences obtained by TIGR. The hope is to identify gene function by matching fragment sequences with the sequences of genes of known function stored in databases. Another goal of TIGR is to analyze the anatomy of gene expression in human organs such as the brain and the liver. Such studies could follow the cascade of gene activation that leads to those cancers caused by oncogenes. TIGR considers a critical technology to be algorithm development allowing the identification of these genes from many others relying only on the portion of the EST whose sequence is known. Tissue-specific cDNA expression is being studied in the hippocampus, fetal brain, and infant brain in order to trace the development of childhood diseases, and perhaps delineate the proteins involved in several childhood neurological diseases.

A number of other commercial companies have also developed novel diagnostics and therapeutics from human genome technology. These include Genomyx and Hyseq (rapid sequencing technology), Mercator Genetics (chromosome mapping), Incyte Pharmaceuticals (cDNA and EST sequencing), Sequana, Inc. and Myriad Genetics, Inc. (diagnostics from gene markers), Darwin Molecular Corp. (large-scale sequencing and directed molecular evolution of targets to T-cell receptor genes), and Millennium Sciences, Inc. (mapping and sequencing of multifactorial gene disorders such as diabetes and hypertension). Myriad Genetics is looking at genes that predispose to common disorders such as breast cancer (a diagnostic kit based on the identification of the recently discovered BRCA1 and BRCA2 genes is under development), melanoma, prostate and colon cancers, heart disease, and hypertension. In addition a test is in development for mutant p53 genes in breast cancer cells along with other cancer diagnostics. Mercator Genetics is looking at still broader genetic illnesses such as cardiovascular disease, some cancers, many psychiatric disorders, and some endocrine disorders. In addition to the first-tier genome companies, a number of service organizations have developed or expanded to assist in the HGI effort. These include companies developing new reagents (BIOS Labs., Inc. has developed kits for coupled DNA amplification), bulk DNA/RNA isolations and characterizations, automated DNA sequencers, and fast chips and dedicated boards for rapid gene comparison (Applied Biosystems Division/Perkin Elmer, Hitachi). One area of intense technological focus is the development of new automated or semiautomated DNA sequencers (ABI, Pharmacia, Genomyx, Hyseq). A related application is the development of high-density arrays of oligonucleotide probes to which genes may be hybridized. A combinatorial chemistry approach to light-directed spatial synthesis of high-density oligonucleotide probes arrayed on a chip has been developed (Affymax/Affymetrics; see Chapter 3 for more details on light-directed spatial synthesis). These arrays have been used to determine the genetic diversity in the reverse transcripase and protease genes of HIV (Lipshutz et al. 1995).

1.6. SUMMARY

The Human Genome Initiative has led to a paradigm shift in modern biology, from the older analog-based "optical" biology, where the biologist observed living processes at ever finer levels of detail, to a newer digital-based biology wherein the ability to collect, sort, analyze, and compare the genetic alphabet is leading to an in-depth understanding of the molecular basis of many living processes. A combination of public collaborative efforts and private enterprise competition has resulted in a variety of strategies for mapping and sequencing the genome—a largely cartographic effort whose scientific and therapeutic benefits are dependent on a considerable number of scientific experiments *after* the sequencing is complete. In particular, therapeutic benefits from the HGI requires the cloning, expression, and isolation of the gene products and their analyses in order to deduce the needed structure-function-information relationships required for novel drug discovery.

The next three chapters describe therapeutic methodologies that will benefit considerably from the information available from the HGI: the use of genes as therapeutic molecules (gene therapy), the use of gene fragments, protein fragments, and small organic molecules combined in a multitude of ways to create new drugs (combinatorial chemistry), and the use of computer-based methods to produce therapeutic proteins or small-molecule drugs that affecting the expression of genes identified by the HGI (protein engineering and computational drug design).

REFERENCES

Adams, M. D., Kerlavage, A. R., Fleischmann, R. D., Fuldner, R. A., Bult, C. J., Lee, N. H., Kirkness, E. E., Weinstock, K. G., Gocayne, J. D., White, O., et al. 1995. Initial assessment of human gene diversity and expression patterns based upon 83 million nucleotides of cDNA sequence. *Nature* 377(6547 Suppl):3–174.

Ashton, R. J., Jaye, M., and Mason, J. S. 1995. *Drug Discovery Today* 1:11–14.

Baptista, R., Kruglyak, L., Xu, S. H., Lander, E., et al. 1995 *Science* 270: 1945–1954.

Barasch, J., Kiss, B., Prince, A., Saiman, L., Gruenert, D., and al-Awqati, Q. 1991. Defective acidification of intracellular organelles in cystic fibrosis. *Nature* 352(6330):70–73.

Beckman, J. S., and Weber, J. L. 1992. Survey of human and rat microsatellites. *Genomics* 12(4):627–631.

Boat, T. F., Welsh, M. J., and A. L. Beaudet. 1989. *The Metabolic Basis of Inherited Disease.* New York: McGraw-Hill, pp. 2649–2680.

Boguski, M. S., Tolstoshev, C. M., and Bassett, D. E., Jr. 1994. Gene discovery in dbEST. *Science* 265(5181):1993–1994.

Bult, C. J., et al. 1995. *Science* 269:496–512.

Caskey, C. T., Kaback, M. M., and Beaudet, A. L. 1990. The American Society of Human Genetics statement on cystic fibrosis screening. *American Journal of Human Genetics* 46(2):393.

Collins, F., and Galas, D. 1993. A new five-year plan for the U.S. *Human Genome Science* 262(5130):43–46.

Collins, F. S. 1991. Of needles and haystacks: Finding human disease genes by positional cloning. *Clinical Research* 39(4):615–623.

Collins, F. S. 1992. Cystic fibrosis: Molecular biology and therapeutic implications. *Science* 256(5058):774–779.

Collins, F. S. 1995. Ahead of schedule and under budget: The Genome project passes its fifth birthday. *Proceedings of the National Academy of Sciences.* 92(24):10821–10823.

Corder, E. H., Saunders, A. M., Strittmatter, W. J., Schmechel, D. E., Gaskell, P. C., Small, G. W., Roses, A. D., Haines, J. L., Pericak-Vance, M. A. 1993. Gene dose of apolipoprotein E type 4 allele and the risk of Alzheimer's disease in late onset. *Science* 261(5123):921–923.

Copeland, N. G., Jenkins, N. A., Gilbert, D. J., Eppig, J. T., Maltais, L. J., Miller, J. C., Dietrich, W. F., Weaver, A., Lincoln, S. E., Steen, R. G., et al. 1993. A genetic linkage map of the mouse: current applications and future prospects. *Science* 262(5130):57–66.

Cotten, M., Langle-Rouault, F., Kirlappos, H., Wagner, E., Mechtler, K., Zenke, M., Beug, H., Birnstiel, M. L. 1990. Transferrin-polycation-mediated introduction of DNA into human leukemic cells: Stimulation by agents that affect the survival of transfected DNA or modulate transferrin receptor levels. *Proceedings of the National Academy of Sciences.* 87(11):4033–4037.

Festing, M. F. W. 1979. *Inbred Strains in Biomedical Research.* Oxford: Oxford University Press.

Fleischmann, R. D., Adams, M. D., White, O., Clayton, R. A., Kirkness, E. F., Kerlavage, A. R., Bult, C. J., Tomb, J. F., Dougherty, B. A., Merrick, J. M., et al. 1995. Whole-genome random sequencing and assembly of *Haemophilus influenzae Rd. Science* 269(5223):496–512.

Fraser, C. M., Gocayne, J. D., White, O., Adams, M. D., Clayton, R. A., Fleischmann, R. D., Bult, C. J., Kerlavage, A. R. Sutton, G., Kelley, J. M., et al. 1995. The minimal gene complement of *Mycoplasma genitalium. Science* 270(5235):397–403.

Goffeau, A. 1994. Yeast. Genes in search of functions. *Nature* 369(6476):101–102.

Guyer, M. S., and Collins, F. S. 1995. How is the Human Genome Project doing, and what have we learned so far? *Proceedings of the National Academy of Sciences,* 92(24):10841–10848.

Harrington, C. R., and Colaco, C. A. 1994. Alzheimer's disease. A glycation connection. *Nature* 370(6487):247–248.

Hudson, T. J., Stein, L. D., Gerety, S. S., Ma, J., Castle, A. B., Silva, J., Slonim, D. K., Baptista, R., Kruglyak, L., Xu S. H., et al. 1995. An STS-based map of the human genome. *Science* 270(5244):1945–1954.

Hyde, S. C., Emsley, P., Hartshorn, M. J., Mimmack, M. M., Gileadi, U., Pearce, S. R., Gallagher, M. P., Gill, D. R., Hubbard, R. E., and Higgins, C. F. 1990. Structural model of ATP-binding proteins associated with cystic fibrosis, multidrug resistance and bacterial transport. *Nature* 346(6282):362–365.

Kerem B., Rommens, J. M., Buchanan, J. A., Markiewicz, D., Cox, T. K., Chakravarti, A., Buchwald, M., and Tsui, L. C. 1989. Identification of the cystic fibrosis gene: Genetic analysis. *Science* 245(4922):1073–1080.

Kosik, K. S. 1992. Alzheimer's disease: A cell biological perspective. *Science* 256(5058):780–783.

Lipshutz, R. J., Morris, D., Chee, M., Hubbell, E., Kozal, M. J., Shah, N., Shen, N., Yang, R., and Fodor, S. P. 1995. Using oligonucleotide probe arrays to access genetic diversity. *Biotechniques* 19(3):442–447.

Leff, S. E., Brannan, C. I., Reed, M. L., Ozcelik, T., Francke, U., Copeland, N. G., and Jenkins, N. A. 1992. Maternal imprinting of the mouse Snrpn gene and conserved linkage homology with the human Prader-Willi syndrome region. *Nature Genetics* 2(4):259–264.

Levy-Lahad, E., Wasco, W., Poorkaj, P., Romano, D. M., Oshima, J., Pettingell, W. H., Yu, C. E., Jondro, P. D., Schmidt, S. D., Wang, K., et al. 1995. Candidate gene for the chromosome 1 familial Alzheimer's disease locus. *Science* 269(5226):973–977; *Science* 269(5226):970–973.

Murray, J. C., Buetow, K. H., Weber, J. L., Ludwigsen, S., Scherpbier-Heddema, T., Manion, F., Quillen, J., Sheffield, V. C., Sunden, S., Duyk, G. M., et al. 1994. A comprehensive human linkage map with centimorgan density. Cooperative Human Linkage Center (CHLC). *Science* 265(5181):2049–2054.

Peltomaki, P., Aaltonen, L. A., Sistonen, P., Pylkkanen, L., Mecklin, J. P., Jarvinen, H., Green, J. S., Jass, J. R., Weber, J. L., Leach, F. S., et al. 1993. Genetic mapping of a locus predisposing to human colorectal cancer. *Science* 260(5109):810–812.

Poustka, A., Pohl, T. M., Barlow, D. P., Frischauf, A. M., and Lehrach, H. 1987. Construction and use of human chromosome jumping libraries from NotI-digested DNA. *Nature* 325(6102):353–353.

Rosenfeld, M. A., Yoshimura, K., Trapnell, B. C., Yoneyama, K., Rosenthal, E. R., Dalemans, W., Fukayama, M., Bargon, J., Stier, L. E., Stratford-Perricaudet, L., et al. *In vivo* transfer of the human cystic fibrosis transmembrane conductance regulator gene to the airway epithelium. *Cell* 68(1):143–155.

Riordan, J. R., Rommens, J. M., Kerem, B., Alon, N., Rozmahel, R., Grzelczak, Z., Zielenski, J., Lok, S., Plavsic, N., Chou, J. L., et al. 1989. Identification of the cystic fibrosis gene: Cloning and characterization of complementary DNA. *Science* 245(4922):1066–1073.

Riordan, J. R., Alon, N., Grzelczak, Z., Dubel, S., and Sun, S. Z. 1991. The CF gene product as a member of a membrane transporter (TM6-NBF) super family. *Advances in Experimental Medicine and Biology* 290:19–20.

Rommens, J. M., Iannuzzi, M. C., Kerem, B., Drumm, M. L., Melmer, G., Dean, M., Rozmahel, R., Cole, I. L., Kennedy, D., Hidaka, N., et al. 1989. Identification of the cystic fibrosis gene: Chromosome walking and jumping. *Science* 245(4922):1059–1065.

Rise, M. L., Frankel, W. N., Coffin, J. M., and Seyfried, T. N. 1991. Genes for epilepsy mapped in the mouse. *Science* 253(5020):669–673.

Smith, M. A., Taneda, S., Richey, P. L., Miyata, S., Yan, S. D., Stern, D., Sayre, L. M., Monnier, V. M., and Perry, G. 1994. Advanced Maillard reaction end products are associated with Alzheimer disease pathology. *Proceedings of the National Academy of Sciences.* 91(12):5710–5714.

Stopa, E., Vlassara, H., Bucala, R., Manogue, K., and Cerami, A. 1994. Advanced glycation end products contribute to amyloidosis in Alzheimer disease. *Proceedings of the National Academy of Sciences.* 91(11):4766–4770.

Terry, R. D., Katzman, R., and Bick, K. L. (eds.). 1994. *Alzheimer Disease*. New York: Raven.

Trapnell, B. C., Yoneyama, K., Rosenthal, E. R., Dalemans, W., Fukayama, M., Bargon, J., Stier, L. E., Stratford-Perricaudet, L., et al. 1992. *In vivo* transfer of the human cystic fibrosis transmembrane conductance regulator gene to the airway epithelium. *Cell* 68(1):143–155.

Waldrop, M. M., 1995. On-line archives let biologists interrogate the genome. *Science* 269(5229):1356–1358.

Williams, N. 1995. Closing in on the complete yeast genome sequence. *Science* 268(5217):1560–1561.

White, R., and Lalouel, J. M. 1988. Sets of linked genetic markers for human chromosomes. *Annual Review of Genetics* 22:259–279.

Vitek, M. P., Bhattacharya, K., Glendening, J. M., Stopa, E., Vlassara, H., Bucala, R., Manogue, K., and Cerami, A. 1994. Advanced glycation end products contribute to amyloidosis in Alzheimer disease. *Proceedings of the National Academy of Sciences.* 91(11):4766–4770.

2

HUMAN GENE THERAPY

2.1. AN OVERVIEW OF MOLECULAR MEDICINE

Up to 4000 human inherited disorders are thought to exist. Although fortunately some of these disorders are rare, others are very prevalent. As many as 1 in 25 people of northern European origin is a carrier of a cystic fibrosis mutation, and the number of people with anemias, muscular dystrophy, hypercholesterolemia, and other such diseases is in the millions.

The leaps made in molecular genetics and recombinant DNA technology have led to the identification of numerous disease-causing genes. The list of diseases for which a candidate gene is identified grows daily, and the Human Genome Initiative (Chapter 1) has been rapidly determining the location of these genes and will no doubt identify many new genes involved in disease. Most diseases are the result of a defect in a single gene, sometimes with only an error in a single base pair. The defect can have extensive, complex, and even fatal effects. Diseases like cystic fibrosis are difficult to treat with conventional drugs, but, a new therapeutic agent—the gene itself—provides hope for the cure of this and other such diseases (for a recent review, see Wolfe 1994).

With the localization and cloning of the defective gene in cystic fibrosis in 1987, the possibility of curing it and other such diseases by the introduction of the correct, nondefective, gene into the patients has been postulated (e.g., see Colledge 1994). This concept, termed *gene therapy,* is not limited to this use; it can be applied to eliminate a pathogenic gene, to replace or supplement a defective gene, or to add a new gene in order to create a more favorable phenotype. Its use has been extended to include the manipulation of cells at the molecular level for cancer therapy (Schmidt-Wolf and Schmidt-Wolf, 1994) as well as the introduction of antigens to elicit an immune response (so-called genetic vaccination).

This chapter will describe current industrial and academic efforts aimed at making gene therapy a reality (see Table 2.1). The difficulty with this approach lies in introducing the gene into the body in a "permanent" fashion, where it can then work in a normal, regulated manner. The ideal inherited disease candidate for gene therapy should satisfy the following criteria:

Table 2.1 Common candidate diseases for gene therapy

Disease	Chromosome Location	Defective Gene
Sickle cell disease	11p	Beta chain of hemoglobin
Cystic fibrosis	7	ATP-binding cassette (ABC) family protein
Huntington's disease	4	Identified; function unknown
Duchenne's muscular dystrophy	X	Dystrophin
Phenylketouria mental retardation	12q	Phenylalanine hydroxylase
Lesch-Nyhan syndrome	Xq	Hypoxanthineguanine-phosphoribosyltransferase
Gaucher's disease	1q	Glucocerebrosidase
Tay-Sach's disease	15q	Alpha chain of lysosomal hexosaminidase A
Galactosemia		Galactose accumulation
Maple syrup urine disease		Amino acids and ketoacid accumulation
Adenosine deaminase (ADA) deficiency	20q 13.11	Adenosine deaminase
Thalassemias	11p	Beta chain of hemoglobin
Alpha1-antitrypsin deficiency	14q	Alpha1-antitrypsin
X-linked chronic granulomatous disease (CGD)	Xp	Cytochrome b558 gp91-phox
Autosomal CGD	7q	P47-phox
Familial hypercholesteremia	19p	Receptor for low-density lipoprotein
Ornithine transcarbamylase deficiency	Xp	Ornithine transcarbamylase
Scid (severe combined immunodeficiency syndrome)	Xq	IL-2 alpha receptor chain
Purine nucleoside phosphorylase deficiency	14q	Purine nucleoside phosphorylase
Leukocyte adhesion deficiency	21q	CD18
X-linked hyper IgM syndrome	Xq	CD40

1. It must be caused by a single gene defect.
2. It affects only one cell type which must be accessible for removal/replacement.
3. The disease causing gene must be already identified and cloned.
4. The regulated expression of this gene is not required to correct the defect.

Gaucher's disease, described in Chapter 1, represents such a disease. The pros and cons of potential delivery systems, and the problems faced in the optimization and regulation of protein expression of genes introduced for therapy *in vivo,* will be considered.

2.2. TECHNOLOGICAL DEVELOPMENTS IN GENE THERAPY

The basic steps in any gene therapy protocol are as follows:

- *Isolation of the gene to be used for the gene therapy.* This may be the normal counterpart of a monogenic disease related gene or an anticancer gene or a gene product to which an immune response would be beneficial to an individual.
- *Delivery and targeting to the appropriate tissue.* Potential vehicles for delivery include normal human cells isolated from the individual, viral-based vector systems, synthetic nonviral delivery systems such as cationic lipids and polymers and "naked" DNA.
- *Controlled expression of the therapeutic gene.* Selective control of gene expression is desired when using gene therapy for vaccination or specifically for targeting cancer cells for destruction. In addition the levels of protein expression in gene therapy may need to be controlled precisely and even temporally in some cases.

2.3. DEVELOPMENT OF DELIVERY SYSTEMS

Cellular Modalities of Gene Insertion

In many cases cells can be transfected with a gene *in vitro* and transferred into a patient. This is referred to as the autologous *ex vivo* approach. The advantages of such an approach are:

1. Cells from the patient will not be confronted and rejected by the immune system (this is a potential problem with viral systems).

2. Transfecting the cells *in vitro* is more efficient, and in some cases it may be possible to isolate specifically the transfected cells with use of selection markers on the vectors.
3. The cell provides an ideal milieu for the DNA/vector containing the therapeutic gene and will protect it from degradation.

For a cell to be used in the above manner it must satisfy some prerequisites. First, it must be easily removed from the body, and it must be able to withstand the traumas of removal and subsequent manipulation, and reimplantation. Second, these cells should be capable of dividing and passing their genetic material to their descendants. Finally, these cells should be able to express a range of gene products. Only a few cell types meet these criteria such as bone-marrow-derived blood cells and skin cells. Bone-marrow cells contain the stem cells that are the precursors for all the cells of the hematopoietic series (this includes red blood cells and all immunological cells such as T cells and B cells and macrophages). Some of these blood cells are long living and circulate throughout the body. Collection and transplantation of bone marrow is now a routine clinical procedure. However, some genes are difficult to express in bone-marrow cells, and bone-marrow transplantation is an expensive procedure. Two types of skin cells can also be utilized for gene therapy. Fibroblasts are the principal cells of connective tissue in the human body. Skin fibroblasts are easily infected by retroviral constructs, grow well in culture, and have been shown to express proteins such as adenine deaminase (ADA, an enzyme found in B cells) and factor IX (a protein involved in blood clot formation), which are not normally produced by these cells. Their reintroduction into a patient by injection under the skin or by skin graft is cheaper than bone-marrow transplantation, although unlike the latter, fibroblasts do not have a depot (cf. bone marrow) to populate, and they do not circulate in the bloodstream.

Skin keratinocytes, the primary cells in the epidermis, are also suitable cells for gene therapy. Cultured cells form sheets that are often used to regenerate skin in burn victims. Liver cells or hepatocytes are also excellent choices for gene carriers in view of the many metabolic defects that affect the liver, such as hemophilia, hypercholesterolemia, and antitrypsin defects. However, adult liver cells are refractory to infection with the types of vectors currently in use, and they are not currently conducive to removal and reimplantation. Some preliminary research has shown that hepatocytes can be cultured long enough to transfect. Reintroduction in these cases has been through embedding and implantation of these cells on a solid support matrix (Matrigel).

A preliminary report has recently described the use of cultured retinal pigment epithelial cells (RPE) to provide gene therapy for retinal degeneration (Dunaief et al. 1995). RPE cells carrying a marker gene (introduced by a retroviral construct; see Section 2.2) were transplanted into the subretinal space of albino RCS rats that have an inherited retinal degeneration syndrome. These cells helped preserve photoreceptors normally destroyed in untreated rats, and the marker gene was expressed in the RPE cells for several months post-transplantation.

In a very early process of development is a yolk sac stem cell derived endothelial miniorgan that can deliver therapeutic proteins. These cells lack expression of the transplantation rejection targets (MHC molecules; see Chapter 5), can self-renew rapidly in an undifferentiated state, and can be induced to differentiate. Thus they are excellent candidates for use in gene therapy protocols for inherited and acquired diseases. When cultured murine yolk sac cells transfected with a growth hormone gene and suspended in Matrigel were subcutaneously injected into experimental mice, the injected cells were shown to form discrete vesicular structures within the Matrigel implant, suggesting directed differentiation of the embryonic yolk sac cells into endothelial tissue. These cells expressed the growth hormone for over four months, proving the utility of these cells for delivery of therapeutic genes (Wei et al. 1995).

Other Modalities of Gene Insertion

Ideally gene therapy would target a defective gene for replacement by a normal one. This would leave the newly introduced gene in its natural site to be controlled by normal regulatory mechanisms. This is not currently feasible, but rapid progress in our understanding of the cellular mechanisms of nuclear import and the identification of novel sequence-specific integrases that mediate the insertion of DNA molecules into specific target sequences may make this possible in the future. Current approaches are focusing on the development of vectors that will introduce specific genes into cells. The novel gene may then integrate into the host genome or remain at an extrachromosomal site, and exert its effect over the defective gene.

Viral Vectors

Retroviral Vectors The unique life cycles of retroviruses and their compact genome (which makes engineering of constructs practically easier) make them ideal vectors for gene therapy. Upon attachment to a specific cell surface receptor, the parent virus enters the susceptible host cell. The viral RNA genome is then copied to DNA by the virally encoded reverse transcriptase which is conveniently carried inside the parent virus. This DNA is transported to the host cell nucleus where it subsequently integrates into the genome; it is then referred to as the *provirus*. The provirus is stable in the chromosome during cell division and is transcribed like other cellular proteins. The provirus encodes the proteins and packaging machinery required to make more virus, which can leave the cell by budding (see Fig. 2.1*a* on page 43).

The retroviral genome is conveniently divided (see Fig. 2.1*b* on page 43); the genes for virion proteins and enzymes (gag, pol, env, etc.) are juxtaposed and flanked at both ends by regions (LTRs) that are responsible for proviral integration and transcription. Encapsidation (packaging of the viral genome into its protein envelope) of the retroviral RNAs occurs by virtue of the psi sequence located at the 5′ end of the viral genome. In recombinant retroviral vectors developed for gene therapy, the entire protein coding regions are removed and replaced by the foreign gene of interest in order to leave a virus capable of integrating its genome but unable to propagate itself due to a lack of structural proteins. To allow propagation and isolation of the recombinant virus, a packaging or helper cell line is utilized (see Fig. 2.2 on page 44). This entails isolation of the retroviral gag/pol and env genes and their introduction separately into a host cell to produce a "packaging line" which will stably produce the proteins required for packaging retroviral DNA but cannot encapsidate due to the lack of a psi region. However, when the recombinant vector DNA carrying the gene of interest is introduced into these cells, the "helper" proteins can package the psi-positive recombinant vector to produce a recombinant virus stock. This can now be used to infect cells to introduce the foreign gene into its genome. The recombinant virus whose genome lacks all genes required to make viral proteins can infect once only and cannot propagate. Hence the foreign gene is introduced into the host cell genome without the generation of potentially harmful retrovirus.

The problems associated with this modality include the following:

1. Since the site of integration is random, it can inactivate a critical gene or activate a harmful one such as an proto-oncogene (a cancer-causing gene).
2. Since retroviruses have an inherently high rate of genetic polymorphism, the potential for generation of inactive viruses exists.
3. Retroviruses infect actively dividing cells only.
4. Retroviruses can be recognized and eliminated rapidly by immune related complement proteins (see below).
5. Elements in the LTR region can affect human gene expression.
6. It is sometimes difficult to generate high-titer recombinant virus because insertion of the foreign gene can cause instability or adversely affect production of vector stocks.
7. The tropism (attraction cr affinity for particular cell types) of retroviral vectors may not correlate to the required specificity.

(a) Retroviral life cycle

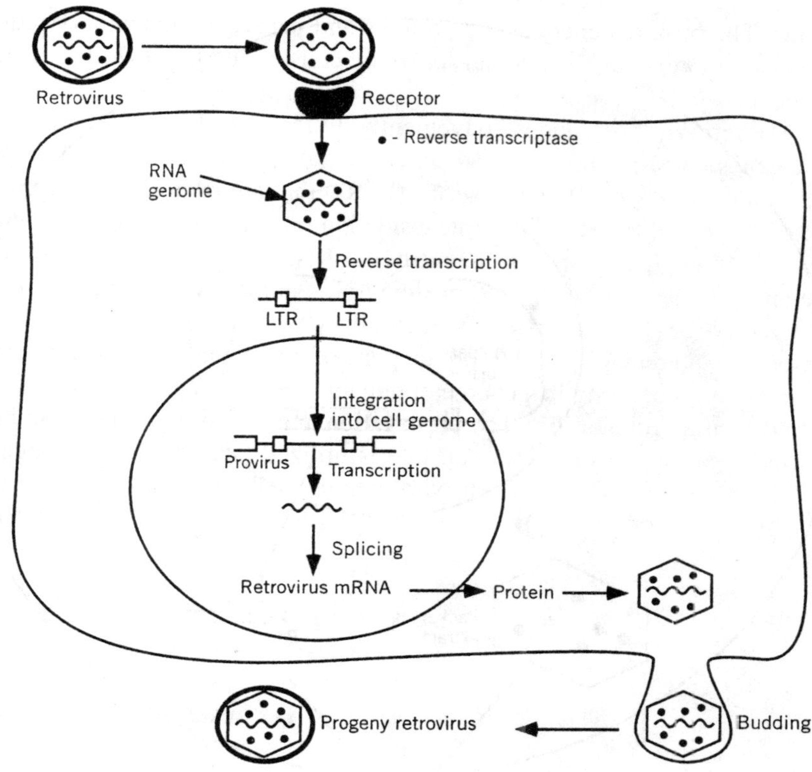

(b) Genome organization

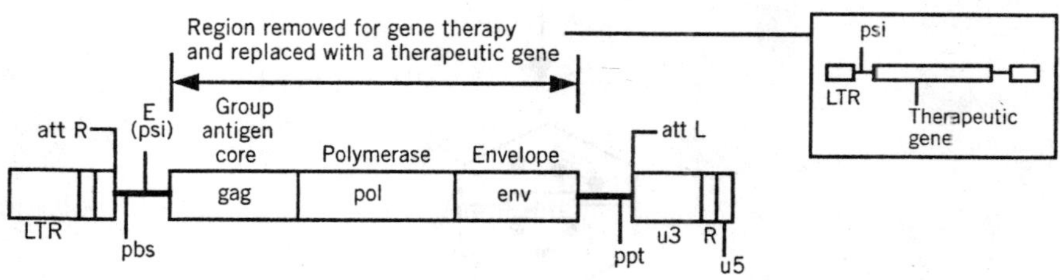

Figure 2.1*a*. The retroviral life cycle. Envelope glycoproteins on the parental virus attach to cellular receptors and accommodate virus entry into the cell. The RNA genome is reverse transcribed to DNA and transported to the nucleus where it integrates into the host genome to form the provirus. This is subsequently transcribed, translated, and the proteins packaged to form a virus particle that leaves the cell by budding with the cell membrane.

Figure 2.1*b*. Genomic organization of a retrovirus. LTR-long terminal repeat; pbs-primer binding site; ppt-polypurine tract; E (psi)-Encapsidation or packaging signal; gag-group specific viral core proteins; pol-polymerase; env-envelope glycoprotein. The solid line represents untranslated regions. The entire gag, pol, and env regions may be replaced by the gene of interest in retroviral vectors (box).

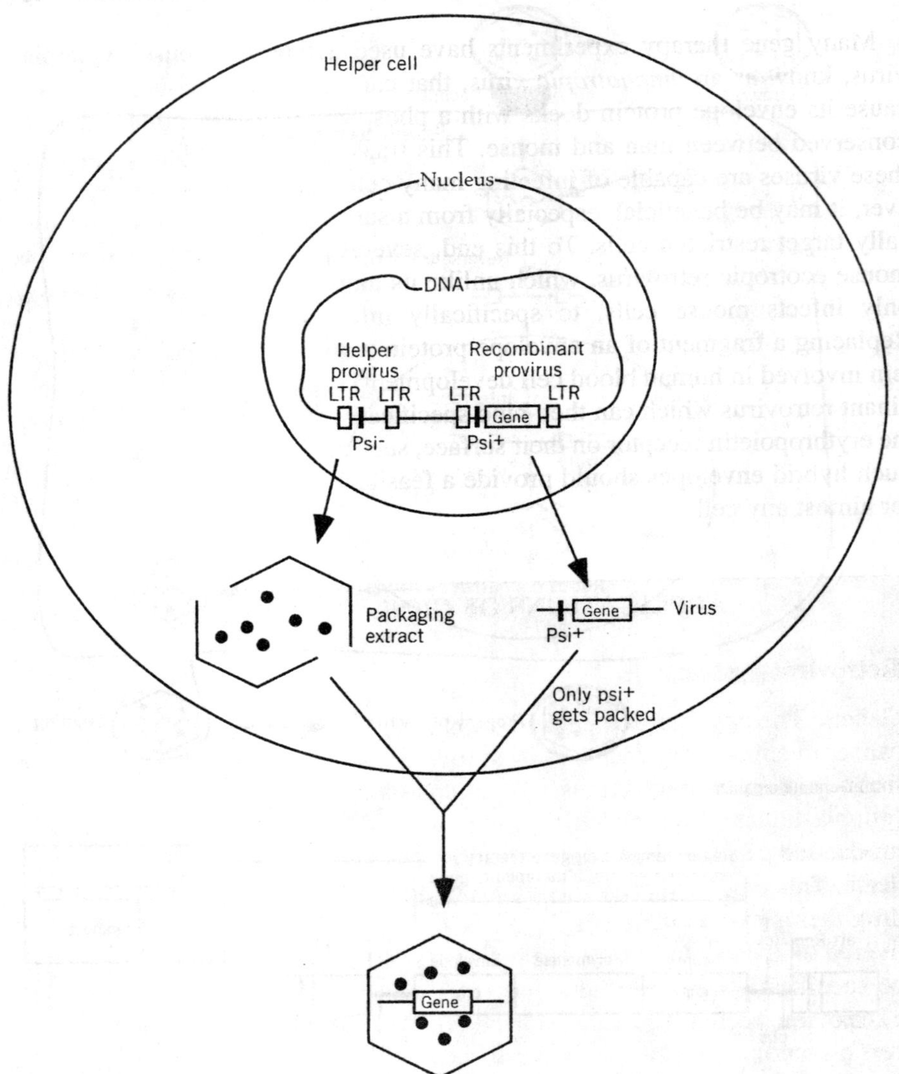

Figure 2.2. Propagation of recombinant retroviruses. A helper virus may be produced by removal of the psi region, for example. This virus produces all of the normal viral proteins but cannot package its own RNA because it lacks the psi element. Introduction of this virus into a host cell produces the helper line. Subsequent introduction of a recombinant retroviral vector will result in the packaging of this psi-positive RNA genome into the proteins provided by the helper cells, to form new recombinant viral particles carrying the therapeutic gene.

Many gene therapy experiments have used a type of mouse leukemia virus, know as an *amphotropic* virus, that can also infect human cells because its envelope protein docks with a phosphate transport protein that is conserved between man and mouse. This transporter is ubiquitous, and so these viruses are capable of infecting many cell types. In some cases, however, it may be beneficial, especially from a safety point of view, to specifically target restricted cells. To this end, several groups have engineered a mouse ecotropic retrovirus, which unlike its amphotropic relative normally only infects mouse cells, to specifically infect particular human cells. Replacing a fragment of an envelope protein with an erythropoietin (a protein involved in human blood cell development) segment produces a recombinant retrovirus which can then bind specifically to human cells expressing the erythropoietin receptor on their surface, such as red blood cell precursors. Such hybrid envelopes should provide a feasible means to attain specificity for almost any cell.

APPLICATIONS OF GENE THERAPY

Retroviral Systems

Genetic Therapy, Inc. has clinical trials using *ex vivo* transduced cells for cancer treatment. They have utilized thymidine kinase (TK) producer cell lines implanted into brains of patients afflicted with brain tumors (glioblastomas). Administration of a drug called gangcyclovir leads to the production of a metabolic toxin specifically in the TK cells and so to cell death. This type of approach is referred to as virally directed enzyme prodrug therapy or VDEPT (Fig. 2.3). In this strategy a foreign gene is delivered to normal and cancerous cells (in some cases the tumor cells may be specifically targeted, but some virus may spread to normal cells) by a retroviral vector. The foreign gene codes for an enzyme that can convert a nontoxic prodrug (e.g., 5-fluorocytosine) to a toxic metabolite (5-fluorouracil) that will kill those cells making it (Sikora et al. 1994). If the promoter utilized is tumor specific, then the toxic product will only be synthesized in the tumor cells. Studies in animal models have demonstrated that this type of treatment can deliver up to 50-fold more drug than by conventional means and has generated a significant rate of complete "cures" in tumors normally resistant to chemotherapy (Connors and Knox 1995). A variation of this technique uses tumor-associated antibodies conjugated to prodrug converting enzymes to provide specific delivery to tumors. This method is referred to as antibody-directed enzyme prodrug therapy (ADEPT).

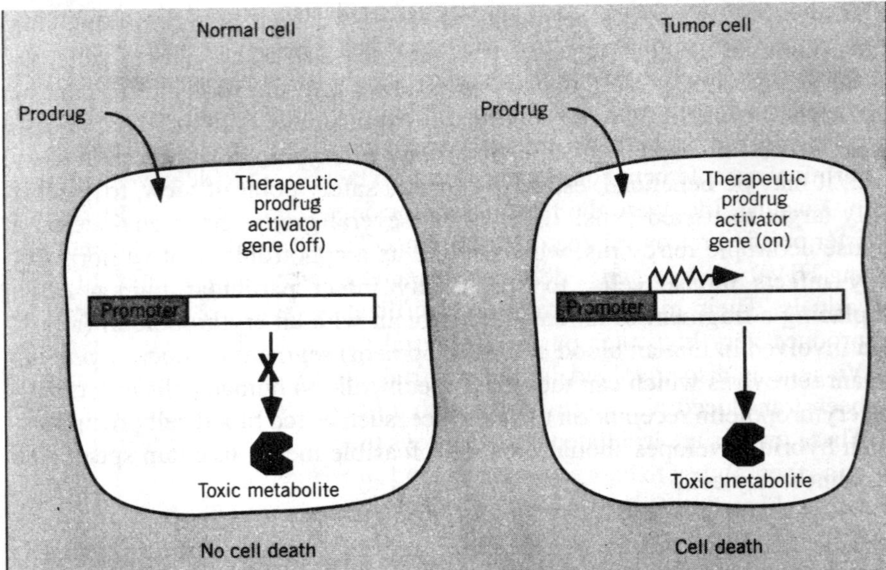

Figure 2.3. Virally directed enzyme prodrug therapy. The foreign gene is linked to a tumor-specific promoter and introduced into cells via retrovirus. This gene encodes an enzyme that converts a prodrug to a toxic metabolic product. The prodrug is nontoxic, whereas its metabolite is cytotoxic. Thus only the tumor cells will produce this and so die selectively.

Cell Genesys, Inc. have developed a novel retroviral packaging system in which high-titer amphotropic retrovirus can be produced without the need to generate stable producer clones. These so-called Kat expression vectors can produce recombinant retrovirus following transient transfection. A high-titer stock is obtained by this method, and both human CD34+ and CD8+ cells have been successfully transduced. The CD34+ population contains the immune stem cells and would thus be ideal for correction of Gaucher's disease, ADA deficiency, or any other disease where an immune cell is defective. These stocks have been demonstrated to be free of replication competent retrovirus by an extensive provirus mobilization assay.

Alexion Pharmaceuticals, Inc. have developed new retroviral particles that target and infect both dividing and nondividing cells. This virus is also engineered to avoid the immune response, particularly the complement cascade. Complement represents a major part of the nonspecific (non–MHC or antibody mediated) arm of the immune response (see Chapter 5). Foreign matter in the form of bacteria or virus can be recognized by certain features such as cell surface carbohydrates by the proteins of the complement system. This recognition leads to the sequential binding of complement and a subsequent destruction of the pathogen. Several findings suggest that the Galα1-3gal glycosidic (sugar-carbohydrate) structure on the surface of retroviral particles is responsible for the complement-mediated inactivation of the virus. Alexion is engineering retroviral vectors deficient in this epitope. Soluble inhibitors that target the terminal complement components can also be utilized to achieve this goal.

Viagene, Inc. were the first and so far only venture to directly administer purified retroviral vectors to patients. They have used vectors encoding HIV-1 env/rev genes delivered intramuscularly to HIV infected individuals. Their manufacturing and purification systems allow them to produce very high titer purified material which the FDA has approved. Viagene is also involved in clinical trials using excised tumor that has been transduced, via a retrovirus, with a human interferon-gamma gene. These tumors are irradiated (this destroys their ability to divide, but maintains their ability to express proteins) and subsequently, returned to the patient where they produce interferon which destroys surrounding tumor cells, but do not proliferate themselves. They too are developing schema to increase the longevity of the recombinant retrovirus by increasing the resistance to serum complement.

Nonviral Systems

Liposome-mediated transfection has been successfully utilized to introduce genes into porcine arteries via balloon angioplasty at low instillation pressures. Here a catheter with a balloon tip is coated with and subsequently used to introduce liposomal encapsulated DNA into blood vessels. Although the subsequent expression is transient, lasting about a month, the potential utility of this modality for treatment of atheroscleriois or other arterial diseases is obvious.

MHC molecules (see Chapter 5) are the targeted proteins in transplantation rejection, and they thus represent targets that initiate rejection. The allogeneic MHC molecule is simply a cell surface protein not normally found in these tumors, and it therefore provides a target for an immune response. Indeed injection of liposomal-HLA-B7 into melanomas not expressing this protein can induce immune response to the HLA-B27, and thus a response to the tumor and eradication of the cancerous nodules. A protocol for immunotherapy of such malignancies by intratumoral injection of an allogeneic MHC gene (HLA-B7) complexed with DC-Chol liposomes was approved by the Recombinant DNA Advisory Committee–NIH (RAC) in 1992. A U.S. phase I study using cationic liposome-mediated transfer of the CFTR gene into the nasal airway is underway. A similar trial has also been approved recently in Great Britain. Vical, Inc. and GeneMedicine, Inc. are companies that are developing lipofection and other nonviral delivery systems.

Polycation Systems

TargeTech, Inc. have developed serum glycoprotein-polycation conjugates that will specifically target the asialoglycoprotein receptor present on hepatocytes in this manner. This method has been used to partially correct inborn errors of metabolism in two animal models and, with antisense strategies, inhibit viral infection in a chronically infected cell line. The antisense approach uses therapeutic genes complementary to the coding sequence, or fragments thereof, of the target gene to downregulate that gene. This inhibition probably occurs by the interaction of the complementary regions leading to a physical blockage in the production of protein from the targeted gene at the level of transcription or translation. Several groups are hoping to target lung epithelial cells through polymerized IgA and surfactant protein B internalization pathways.

Adenoviral Vectors (Kotin 1994) The adenovirus is a double-stranded, linear DNA virus that is the etiologic agent responsible for the common cold and other respiratory ailments. Adenovirus enters the cell by a cell surface receptor mediated mechanism like that used by retrovirus. However, once inside the cell, adenovirus does not integrate into the host genome. Instead it functions episomally (independently from the host genome) as a linear genome in the host nucleus. Hence the use of recombinant adenovirus alleviates the problems associated with random integration into the genome but results in only a transient expression that is not passed onto subsequent progeny. Since adenoviruses demonstrate tropism for a wide variety of tissues, including respiratory airway epithelial cells, hepatocytes (liver cells), muscle cells, cardiac myocytes (heart), synoviocytes (joint tissue), primary mammary epithelial cells, and neurons (nervous system and brain), they can be used for gene therapy to target any of these cells. The liver is the site of many biochemical deficiencies, muscle is the site of muscular dystrophy, and synoviocytes can be targeted for arthritis. Adenovirus also infects nondividing cells. This is very important for a disease such as cystic fibrosis in which the affected cells, the lung epithelium, have a slow turnover rate. In fact several trials are underway utilizing adenovirus-mediated transfer of cystic fibrosis transporter (CFTR) into the lungs of afflicted adult cystic fibrosis patients. Preliminary data indicate, however, that adenoviruses are very immunogenic, and thus a new generation of an adenovirus that does not provoke strong cytotoxic T-cell responses is being developed. The strategy for recombinant adenoviral construction is essentially as described for the retrovirus (Fig. 2.4). Briefly, particular regions of the genome (E1a and E1b) that are essential for replication are replaced by a cassette containing the gene of interest. This vector is propagated in a "helper" line that provides the E1 protein to produce a replication-deficient recombinant virus.

Other Viral Vectors Other viral vectors that circumvent some of the disadvantages of adenoviral and retroviral vectors are currently being developed. A plasmid-based self-amplifying Sindbis virus vector, in which a recombinant cDNA genome of this (+)-strand virus was placed under the transcriptional control of a Rous sarcoma virus (RSV) LTR promoter, produced expression levels of a marker protein that were 200-fold higher than those obtained with a conventional RSV vector (Herweijer et al. 1995). Although the expression was transient, lasting approximately two weeks,

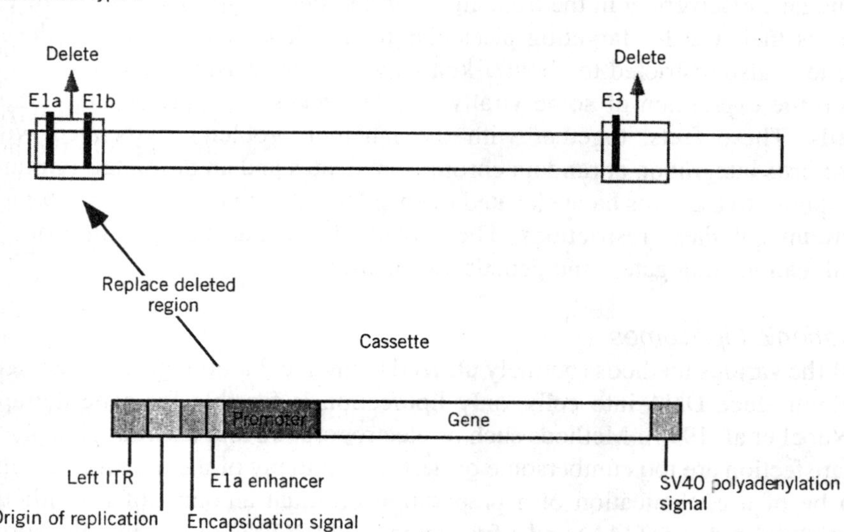

Figure 2.4. Generation of adenoviral vectors. Replication deficient adenovirus vectors are produced by the deletion of the E1 and E3 regions. A cassette containing the therapeutic gene juxtaposed by appropriate regulatory elements can be inserted into the E1 site and recombinant virus prepared utilizing helper lines (see retroviral propagation for analogous details). ITR-inverse terminal repeat, SV40-Simian virus 40.

this vector may allow for safe, short-term delivery of gene products in gene therapy protocols.

Herpes simplex virus type 1 (HSV) is a neurotropic virus that naturally establishes latency in neurons of the peripheral nervous system. Replication defective HSV vectors have been developed; these are deleted for at least one essential immediate early regulatory gene, rendering the virus less cytotoxic, incapable of reactivation but still capable of establishing latency (Glorioso et al. 1995). Development of latency-based promoter systems will allow this vector to potentially target the nervous system for gene therapy.

A parvovirus vector (Targeted Genetics, Inc.) is currently being used to introduce the CFTR into the sinuses of cystic fibrosis patients at the Stanford University Medical Center. The virus vector is administered to the maxillary sinuses (below the eye), which are small, easily accessible, and have the same surface tissue as the lungs. This vector is reported to cause less respiratory inflammation, and a lower immune response than the previously used adenoviral vectors. Success with the sinus route will lead to efforts to target the lungs at a later stage.

Nonviral Systems

The lack of diversity in the tropism of viral systems, especially retroviruses, limits their use for targeting particular tissues. The size of the introduced gene is also restricted to about 7 kbases with current viral vectors. In addition the expression of some virally shuttled genes is suppressed in certain cells. These facts, together with the inherent problems associated with provirus integration at random chromosomal sites and an unwanted immune response to the virus have elevated the need to develop other methods which circumvent these restrictions. These methods include cationic liposomes, polycation conjugates, and genetic vaccination.

Cationic Liposomes

Of the various methods routinely utilized by molecular biologists to transfect or introduce DNA into cells, only lipofection is feasible for gene therapy (Nabel et al. 1993). Methods such as electroporation and calcium phosphate transfection are too cumbersome or destroy too many of the transfected cells to be of use. Sonication of a preparation of equal amounts of a synthetic cationic lipid (DOTMA) and a fusogenic lipid (DOPE) results in a liposome preparation commercially sold as Lipofectin™. Other mixtures can be combined to obtain liposomes or cytosomes. These liposomes are composed of small unilamellar vesicles with a size range of 50–200 nm in diameter. Mixing of these cationic liposomes with an aqueous solution of DNA results in interaction of the two components to form a positively charged complex, containing some encased DNA, that binds and then fuses with the negatively charged cell surface and is subsequently taken up by the cells with high efficiency. The released DNA can then be expressed extrachromosomally within the recipient cell. This system can also be utilized to deliver recombinant proteins, and antisense oligonucleotides into the cell. Several features make this system appealing for introducing genes into mammalian cells for gene therapy:

1. Commercial availability and stability of liposomes.
2. Ease of bulk synthesis of novel liposomes.
3. High binding efficiency for naked DNA or RNA with no limitations on size.
4. Ability to transfect most cells.
5. Lack of immunogenicity or biohazardous activity making repeated usage a possibility.
6. Replication of the recipient cell not necessary for DNA uptake.

Polycation Conjugates

Liposomes fuse nonspecifically to most target cells and thus lack the specificity of retroviral transfer. A cell-specific pathway of nonviral gene delivery would circumvent the problems and limitations associated with viral systems, while maintaining the advantages conferred by liposome-type transfer. Receptor-mediated endocytosis of molecular conjugate vectors represents such a strategy (see Fig. 2.5.). Here a bifunctional molecular conjugate is employed to bind DNA and transport it into the target cell via receptor binding and subsequent internalization by the process of endocytosis (Perales et al. 1994). One-half of the conjugate consists of a DNA binding moiety, such as cationic polylysine, that is covalently linked to a ligand, such as transferrin, for a specific receptor on the target cell. The DNA of interest interacts electrostatically with the polylysine domain and condenses into a compact toroidal (doughnut-shape) structure. The attached ligand can bind its cognate ligand, and the physiologic process of receptor-mediated endocytosis internalizes the entire conjugate into the cell. Inside the cell the complex remains in a vesicle that fuses with another vesicular organelle designated the endosome. Escape from this compartment is required before the DNA can relocate to the nucleus for gene expression.

One of the problems with this mode of delivery is the inability of the transmitted DNA to escape from the endosome. Paradoxically this egress can

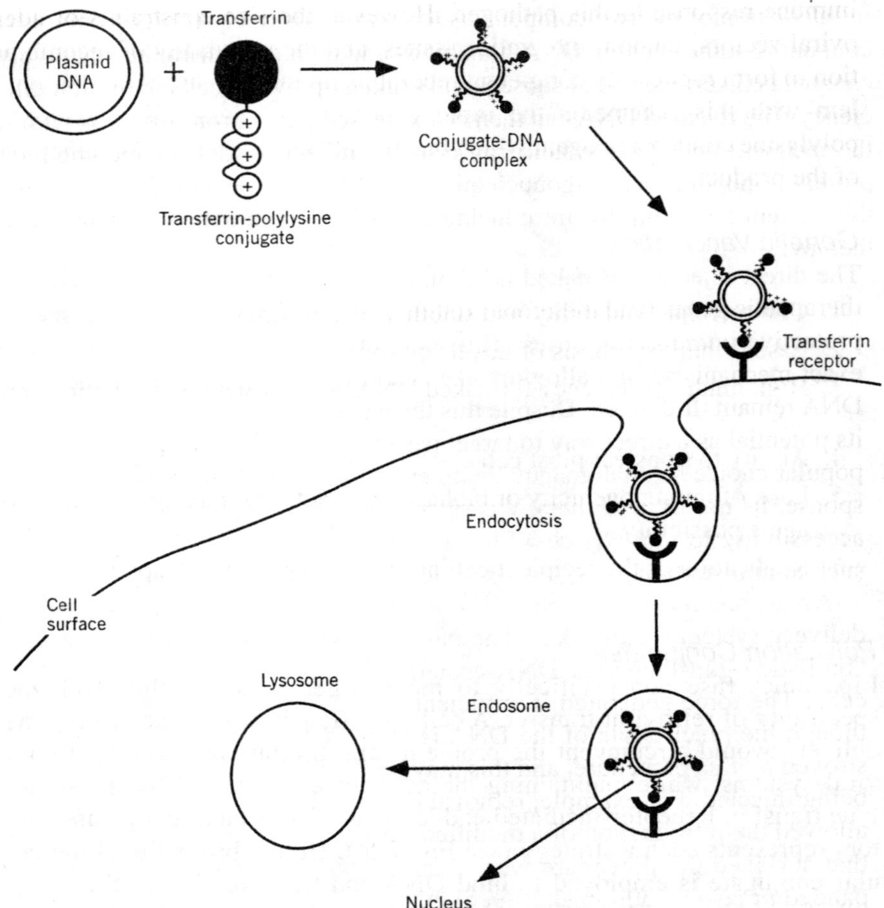

Figure 2.5. Gene transfer by receptor-mediated endocytosis. A bifunctional conjugate consisting of a DNA-binding polylysine moiety and a cell surface receptor ligand (in this case-transferrin) is employed. The ligand-DNA complex binds the appropriate receptor and is internalized by the receptor-mediated endocytic pathway, thus transporting the therapeutic gene into the cell. A fraction of DNA is localized to the nucleus, by undefined mechanisms, where gene expression is effected.

be facilitated by the presence of the adenovirus which has novel mechanisms that allow active endosomal exit. In particular, the hexon protein, with other capsid proteins, is known to aid escape. Strategies that use coadministration of the adenovirus or engineering of adenovirus-polylysine conjugates are also being pursued. The use of the adenovirus brings with it the problems of immune response to this pathogen. However, the size constraints of adenoviral vectors, random integration issues, and the possibility of recombination to form replication competent recombinant are obviated. A further problem with this scheme is the rapid generation of antibodies against the polylysine condensing agent which thereby prevents repeated administration of the product.

Genetic Vaccination

The direct injection of naked DNA into tissues provides a means to deliver therapeutic genes (and therefore proteins) as well as a means to elicit specific protective immune responses. This modality is conceptually crude, and the exact mechanisms that allow uptake, stability, and expression of the vector DNA remain ill-defined. Despite this the technical ease of the technique, and its potential as a direct way to target particular tumor masses, have made it a popular choice for malignancy treatment, mostly for eliciting an immune response. Its use in gene therapy is restricted by its lack of specificity and inaccessibility to most tissues. The skin and muscle are easily accessible to such administration.

Agracetus, Inc. has developed an Accell R particle mediated nucleic acid delivery system for the skin. The basic procedure involved is the use of a "biolistic" gun that fires DNA-coated tungsten or gold microparticles at cells. The force generated is sufficient to drive the DNA into the cells, although the precise fate of the DNA is as yet unknown. Many cells are destroyed by this procedure, and this may be a problem depending on the cells being injected. For example, removal of T cells from HIV-infected patients allowed the introduction of a modified rev gene into these cells. Those cells that survived the procedure were then introduced into the patient and expanded *in vivo* by administration of cytokines. The mutated rev can act in a dominant manner to inhibit rev function in a HIV-infected cell. The HIV virus cannot perform the essential function of transporting its genetic messages from the nucleus to the cytoplasm without rev and thus cannot replicate. Expansion of the mutant rev-carrying T cells would theoretically eradicate the HIV virus in an individual over a period of time. Other genes that target the HIV virus, or any other virus for that matter, may be introduced in this way for antiviral gene therapy. Such genes may be dominant mutant alleles, single-chain antibody genes, antisense genes, or ribozymes. The latter two are preferable from the point of view that since they are RNA based, they will not trigger an undesired immune response. They may also be used in gene therapy strategies for cancer; an anti-p53 ribozyme (catalytic RNA) designed to cleave the p53 (an anti-oncogene that is found in a mutated form in many tumors) pre-mRNA was shown to efficiently reduce the level of mutant p53 mRNA, and it has significantly suppressed the growth of tumor cells in culture (Cai et al. 1995).

Other ventures (GeneMedicine, Inc., Apollen, Inc., Vical, Inc.) and a number of academic laboratories are using genetic vaccination to drive an immune response to tumor antigens such as carcinoembryonic antigen (CEA) and SV-40 large tumor antigen and to pathogenic agents such as herpes simplex, rabies, hepatitis B, and malaria.

2.4. TARGETING AND CONTROL OF GENE EXPRESSION

Cellular targeting of the gene of interest can be achieved through the use of specific receptor-mediated endocytosis-based systems, by direct administration of the gene of interest to the required site by DNA injection, or by balloon angioplasty. Cellular targeting may also be achieved by transfection *ex vivo* of the targeted tissue such as is the case for bone-marrow or by specific induction of expression of the gene of interest in the required target cells. This could be achieved through knowledge and use of tissue-specific promotors in the DNA construct.

Some forms of gene therapy may also require the control of gene expression in either a temporal or transient fashion (Fig. 2.6). Retroviral vectors that contain the tetracycline-inducible (tet) system have been developed (Paulus et al. 1996). Here the gene of interest can be turned on and off by administration of tetracycline, which is nontoxic and has desired pharmacologic properties for this purpose. This system is described in greater detail in the sidebar on generation of mouse models and gene switches.

Steroid receptors, once bound by their cognate ligands, are activated and transported to the nucleus where they can selectively induce the transcription of specific genes. GeneMedicine, Inc. have taken advantage of this paradigm to create a gene switch for use in gene therapy. In their system a drug called antiprogestin activates a recombinant steroid receptor that in turn binds selectively to the promotor region of the therapeutic gene. This gene switch does not affect other cellular pathways, and *in vivo* experiments have demonstrated that the gene switch is activated by orally administered antiprogestin at does 100- to 1000-fold lower than currently indicated for certain clinical applications. Similarly several radiation sensitive genes have been identified and their promotors isolated. Linkage of one such promotor, Egr-1, to the TNF-α (an anti-tumor cytokine) and subsequent delivery of this chimeric gene into tumor cells results in eradication of the tumor upon radiation treatment (as TNF-α production is turned on) at a rate that is better than gene therapy or radiation therapy alone.

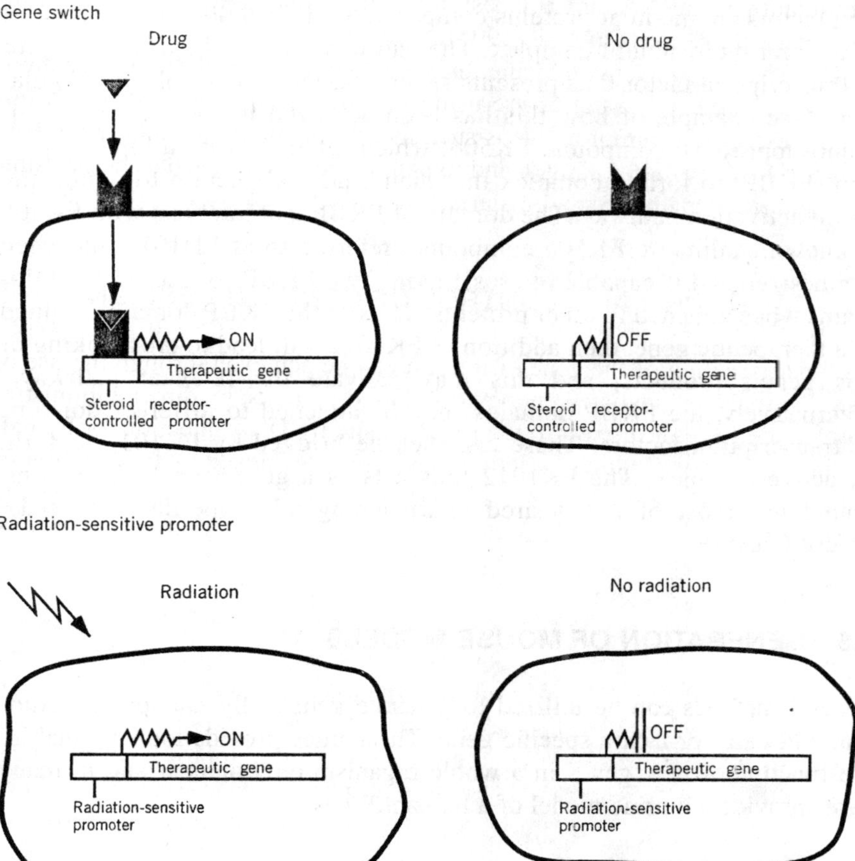

Figure 2.6. Methods to regulate the expression of gene therapeutics. Promoters that are regulated under certain conditions are the key to this strategy. (1). A drug that binds to a steroid receptor to form a complex that can activate a juxtaposed therapeutic gene. In the absence of drug, the receptor does not migrate to the nucleus to activate transcription. The drug used should have the desired pharmacological properties. (2). If the promoter used is radiation sensitive, then this may be used to switch the gene on and off.

Many intracellular processes rely on protein–protein interactions for function. The crosslinking of membrane receptors (naturally by ligands or artificially by antibodies) such as CD3 proteins leads to activation of signal transduction pathways, and many of the steps in such pathways rely on the interaction of two or more intermediates. In addition many transcription factors are modular proteins composed of distinct domains or are part of a larger multisubunit complex. The pathway from cell surface receptor to transcription factor thus presents several targets to control gene regulation. One example of how this has been achieved is the use of the immunosuppressive compound FK506, which interacts with a binding protein (FKBP) to form a complex that then binds calcineurin to inhibit the T-cell activation pathway. The domain of FKBP that interacts with FK506 is known. A dimeric FK506 compound, referred to as FD1012, has been synthesized and is capable of crosslinking two FKBPs or the relevant domains when attached to other proteins. Thus, if the FKBP domain is joined to a therapeutic gene, then addition of FK1012 will lead to crosslinking of this gene's products, and this may activate the relevant pathways. Alternatively, the FKBP domains may be attached to different domains of transcription factors. These can then be bridged by FK1012 to form an active complex. The FK1012 thus acts as a gene switch. This compound has most of the desired pharmacological properties that make its use feasible.

2.5 GENERATION OF MOUSE MODELS

Several methods can be utilized to generate genetically manipulated mice that will carry or lack a specific gene. These mice provide tools to analyze the function of new genes in a whole organism environment and in many cases provide a murine model of a human disease.

Transgenic Mice

Transgenic mice overexpress a gene of interest in a particular tissue (Fig. 2.7). The cDNA (or gene for this protein) is engineered to be preceded by a promotor region, which will direct its tissue-specific expression and contain all other signals required for effective processing of mRNA and final protein expression. Injection of this construct into the pronuclei of fertilized mouse eggs leads to the integration of this DNA into the mouse genome. This event is at a random site and occurs by an, as yet, unclear mechanism. In addition the integration is often by tandem repeats of the construct.

The injected eggs are transplanted into a pseudopregnant female. The resultant pups can be screened for integration of the gene by southern blot analysis of DNA isolated from a small portion of the tail. Subsequently positive pups can be bred to test for Mendelian transmission of the gene or can be bred to homozygosity. Expression of the gene and any resultant phenotypic changes can then be studied.

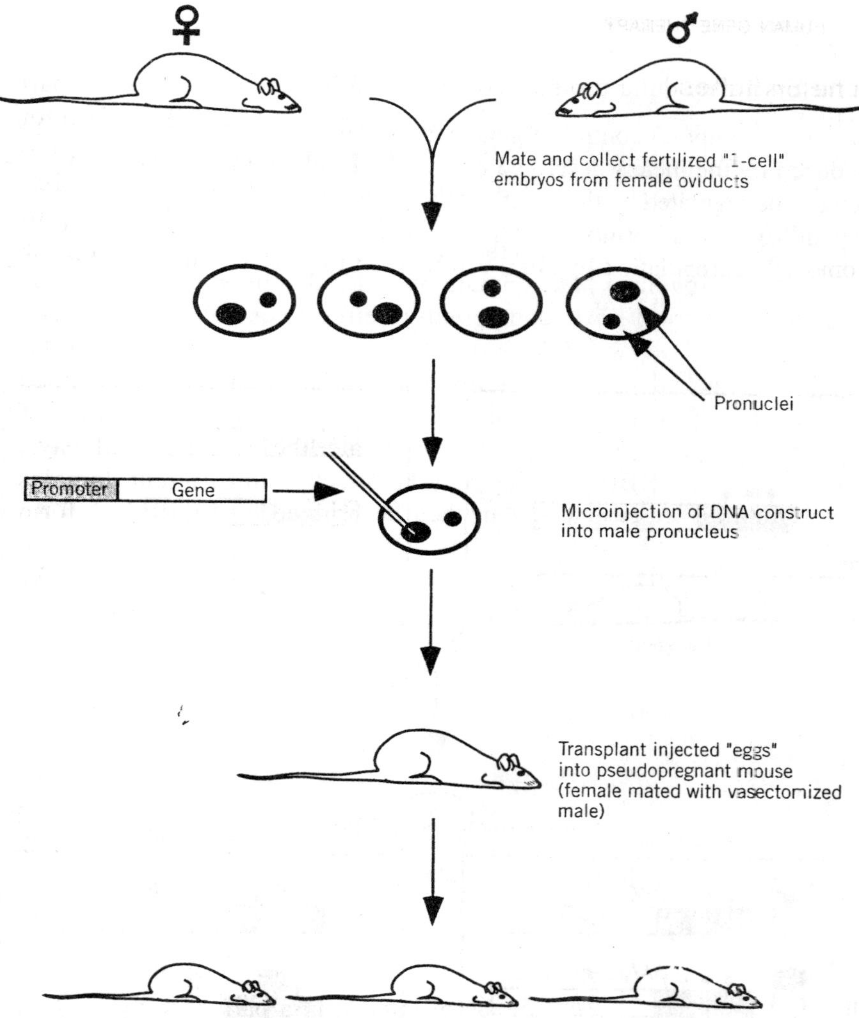

Figure 2.7. Generation of transgenic mice. Superovulation is induced in female mice by injecting intraperitoneally with Pregnant Mare Serum. This is administered three days prior to microinjection. Two days later these mice are injected with Human Chorionic Gonadotrophin and immediately placed with males to "mate" overnight. Female mice that have mated are identified by the presence of a "vaginal plug,' sacrificed, and their oviducts collected. "One-cell" embryos are collected and microinjected with the DNA of interest, free of vector sequences. These embryos are subsequently implanted into "pseudopregnant" females produced by mating with vasectomized males. Any progeny can then be tested for transgene integration, which is frequently performed by analysis of genomic DNA isolated from a small piece of the tail.

Gene Switches

To achieve temporal control of gene expression in transgenic mice (Fig. 2.8) produced by the means described above, the levels of the transgene expression can be regulated by the nature of the promoter used. Thus promoters that respond to steroid hormones or heavy metal ions (e.g., the metallothionein promoter's upregulation by zinc) can be used to produce mice in which the

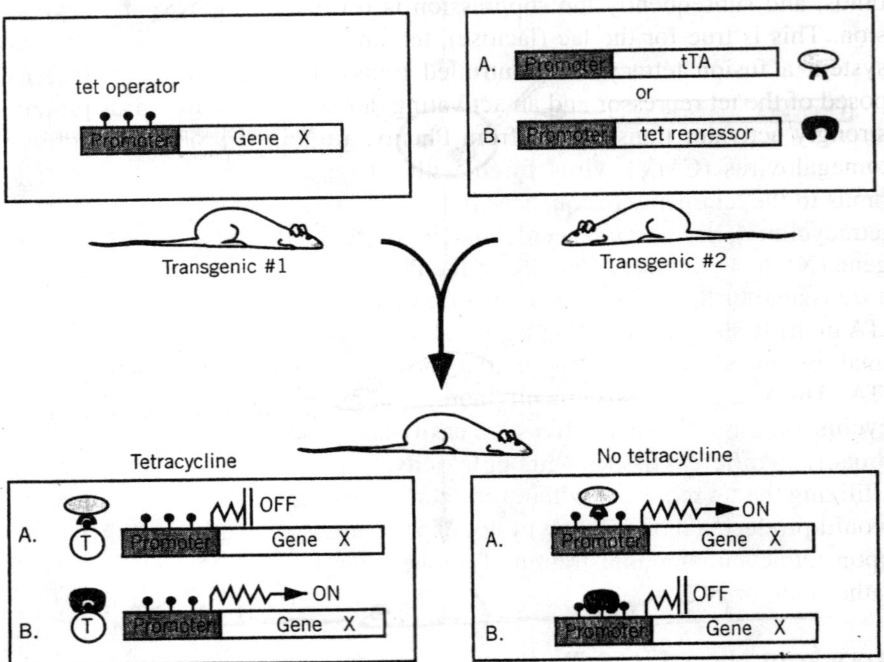

Figure 2.8. Temporal regulation of gene expression in transgenic mice. Two transgenic (TG) mice are made. TG#1 carries the gene to be regulated (X) under the control of a chimeric promoter composed of a minimal cytomegalovirus promoter and tet operator elements. TG#2A carries an activator chimeric protein gene (transactivator-tTA) under the control of a promoter that directs expression in the tissue in which the regulation of gene X is desired. The tTA protein only binds the TG#1 promoter (and therefore activates transcription and expression) in the presence of tetracyline. Thus the progeny of mice produced by crossing TG#1 and TG#2A (double TG mice) will express protein X only in the absence of tetracyline.

Alternatively, a TG mouse (TG#2B) driving the expression of a tet repressor protein can be made. This repressor binds the tet operator only in the absence of tetracycline. Thus TG#1 X TG#2B progeny mice can be induced to express protein X by administering tetracycline.

transgene expression can be modulated. However, toxic effects of the inducers, and the high basal activity of these promoters limit their use. In addition, as with using tissue-specific promoters, the levels of transgene expression are controlled by the levels of endogenous inducers. To enable absolute control over transgene expression, a method utilizing well-characterized regulatory elements from the tet operon of *E. coli* bacteria has recently been described. In bacterial operons a repressor protein binds to a so-called operator to sterically block RNA polymerase progression and thus prevents gene expression. Upon binding by an inducer, however, tis repressor no longer binds, and subsequently the suppression is reversed to allow gene expression. This is true for the lac (lactose), tet, and many other operons. In this system a fusion tetracycline-controlled transactivator protein (tTA) composed of the tet repressor and an activating domain of a herpes virus protein strongly activates transcription from Phcmv, a minimal promoter from cytomegalovirus (CMV), virus fused with tet operator sequences. The tTA binds to the tet operator sequences in the absence but not in the presence of tetracycline. Thus two transgenic mice must be created. The first carries the gene (X) to be regulated attached to the Phcmv promoter; the second carries a transgene of the tTA under regulation of a promoter that will express the tTA in the tissues requiring the temporal modulation. A double transgenic animal, produced by simply mating the above mice, will express X driven by tTA. This can then be specifically abrogated by the administration of tetracycline. Tetracycline derivatives are easily absorbed in mice, are distributed broadly to different tissues without toxicity, and are available commercially. Utilizing the tet repressor without the activator domain in the above scheme would produce a mouse that will not express any X normally but will do so upon tetracycline administration. Thus here the system is used to turn on, rather than off, the gene.

Knockout Mice (Fig. 2.9)

Knockout mice represent the most definitive way to establish the function of a gene *in vivo.* In this method mouse cells are manipulated to generate a mouse in which a gene of choice is selectively inactivated while the remaining genome is unaffected. The technique relies on a phenomenon referred to as *homologous recombination;* if a gene is introduced into a cell line, and it contains sequences that are homologous to an endogenous gene, then the incoming gene can displace the endogenous gene. Thus, in generating knockout mice, a construct containing an inactivated gene of choice (by point mutation or insertion of a neomycin resistance gene into a critical exon) still containing homologous sequences is introduced into embryonic stem (ES) cells. Those cells that contain a disrupted gene can be isolated by a drug re-

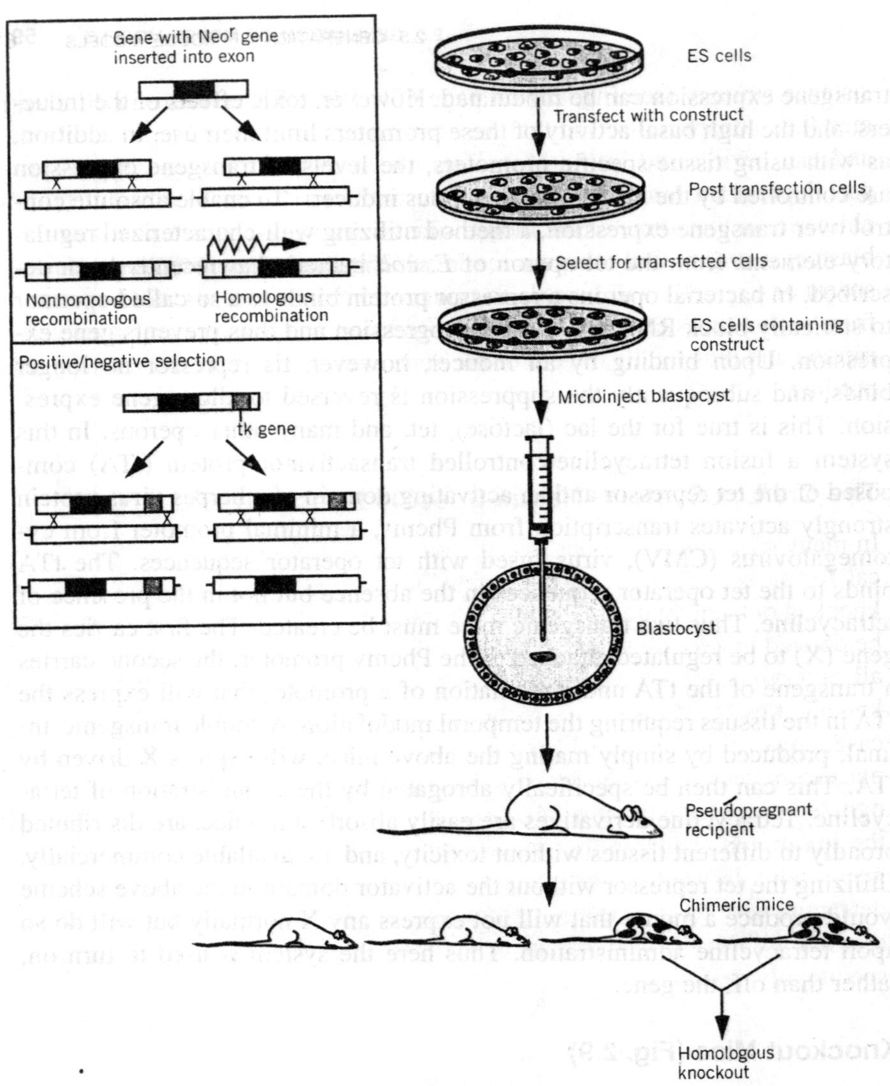

Figure 2.9. Generation of "knockout mice" by homologous recombination. Shown in the box are two strategies for enrichment of homologous recombination events. The gene to be "knocked out" is interrupted by insertion of a Neomycin resistance gene. Transfection of this construct into ES cells can then lead to homologous recombination or random integration of the inactivated gene (*top box*). The required homologous recombinant can be selected by culturing the cells in Neomycin; only in this case can the Neomycin gene be expressed by the promoter of the "knocked-out" gene.

Alternatively, a positive-negative strategy can be utilized (*bottom box*). Here the inactivated gene is placed in a vector containing a juxtaposed thymidine kinase (TK) gene. The vector is introduced into the ES cells, and they are grown in media containing Neomycin and FIAU, a drug that is metabolized by TK to generate a lethal

sistance strategy (see Fig. 2.9 on page 60). These ES cells are undifferentiated (i.e., nonspecialized so that they can go on to develop into any cell type) and can be injected into a blatocyst, which in turn is transplanted into a pseudopregnant female to develop to term. The subsequent progeny will be chimeric, in that some of the mouse's cells will be of ES origin and some of blastocyst origin. Only those mice in which some of these ES cells formed sperm or egg cells will be able to transmit the disrupted gene. The progeny from such a cross can then be bred to produce mice homozygous for the disruption and thus totally devoid of the gene of choice and therefore its protein product.

The Cre/Lox System–Tissue-Specific Knockout (Fig. 2.10)

In many cases the knockout strategy outlined above will delete a gene critical for early embryonic development. In such cases the pups carrying the knocked out gene will die *in utero*. To circumvent this problem, the gene can be deleted in a tissue-specific manner whereby it is less likely to affect overall embryonic development. The first step of the strategy is essentially a knockout by the homologous recombination method described above. In this case, however, a functional gene flanked by lox sequences is used as the displacement vector. The lox sequences are phage derived and are normally recognized by a phage protein called *cre,* which removes all sequences between the lox motifs. Thus, in the second step, a transgenic mouse carrying a cre gene under the control of a promotor of choice is made. This promotor will determine the tissue expressing cre and therefore, when the transgenic is crossed with the first knockout, will determine in which tissue the gene is specifically deleted.

product. Thus cells in which random integration has occurred (*bottom left*) will be sensitive to FIAU, whereas cells in which homologous integration has occurred (*bottom right*) will be resistant to both drugs.

The selected recombinant cells are injected into a blastocyst and implanted into a pseudopregnant mother (see transgenic section). Since the ES cells injected into the blastocyst are usually derived from a different colored strain to that from which the blastocyst is obtained, the resulting progeny should form a chimeric colour. These can then be intercrossed to obtain mice lacking an active gene on both chromosomes.

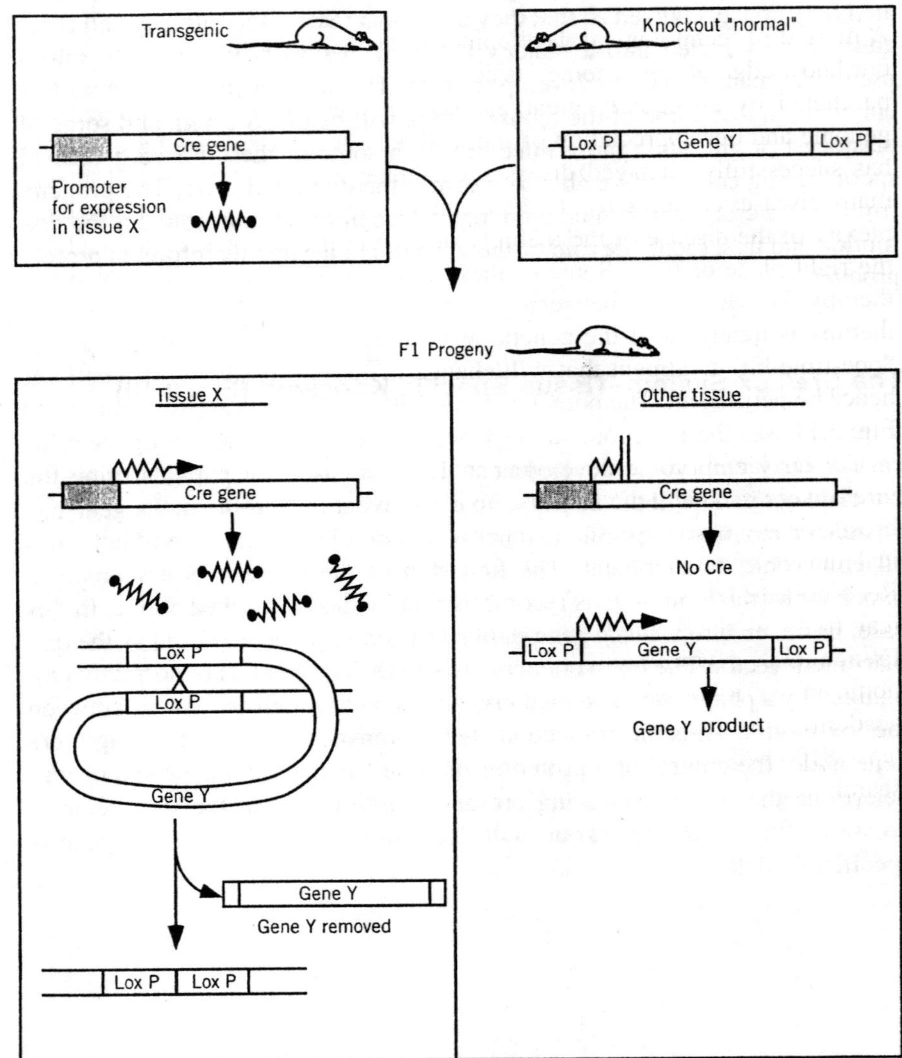

Figure 2.10. The cre/lox system. This strategy allows the tissue-specific inactivation or knockout of genes. Two mice are generated. A TG mice expressing the cre gene under the control of a tissue-specific (where the knockout is desired) promoter and a mouse in which a gene (Y) has been replaced via homologous recombination by gene Y flanked by lox P sites. The progeny obtained upon crossing these mice will express cre protein in the chosen tissue X only, and this will subsequently result in the removal of gene Y in a tissue-specific manner.

2.6. SUMMARY

A myriad of technological developments has fostered an incredible leap in our knowledge of the genome, its contents, and its regulation. This has been paralleled by an equally great increase in our ability to manipulate the genome and our appreciation of intracellular function. Traditional medicine has successfully managed diseases with pharmaceutical agents. However, many diseases are refractive to these conventional treatments due to the complexity of the disease or the inability of current drugs to target the defect at the right place or time. Some of these diseases may well succumb to gene therapy. The difference between gene therapy and drug therapy is that gene therapy is treatment at the genetic level and thus has the potential to be a "one-time-hit" treatment that will change the actual genotype of a cell and hence be permanent. The potential modes of gene therapy are summarized in Fig. 2.11.

Current targets for gene therapy include single gene defects, cancer, and viral diseases. That gene therapy, be it for cancer, atherosclerosis, cystic fibrosis, or any other disease, is to be an important part of scientific and medical development over the next decade is borne out by the fact that the Recombinant DNA Advisory Committee (RAC) has approved over 50 clinical trials in the United States alone as of January 1995, and that the French–U.S. pharmaceutical company Rhone-Poulenc Rorer (RPR) announced that it has set up a network of 14 partnerships with companies and academic institutes working on gene therapy; the company will spend at least $100 million per year over the next several years supporting this research.

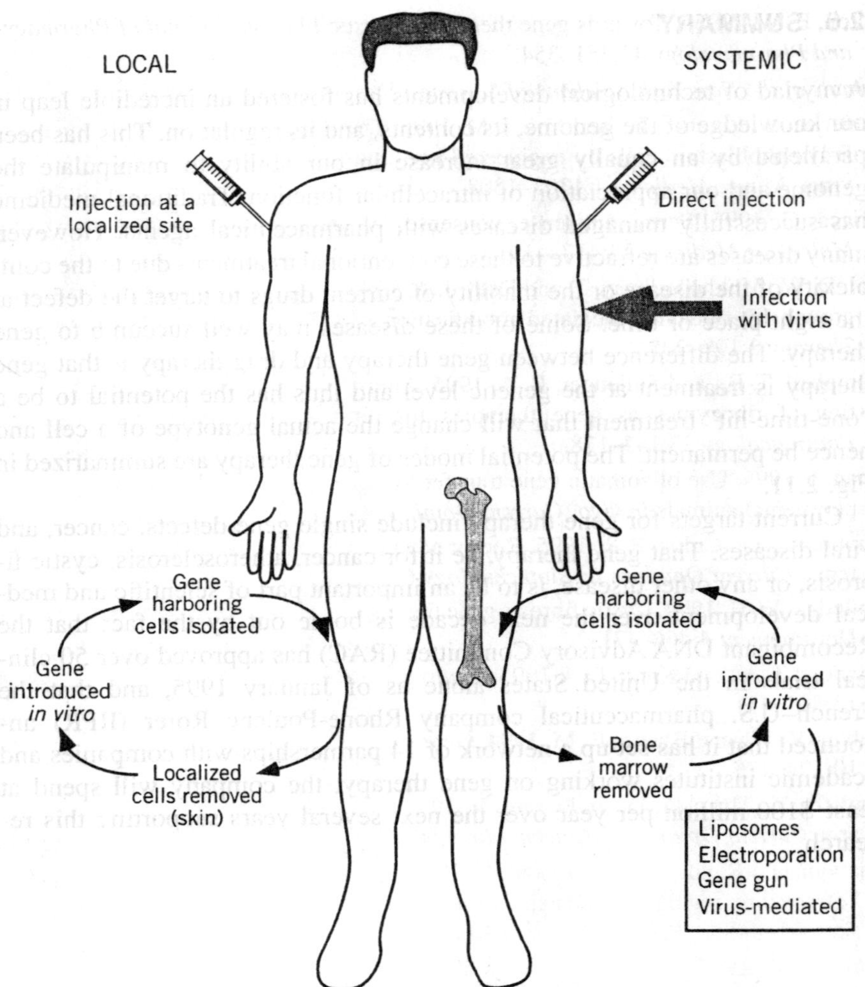

Figure 2.11. Gene therapy options. Gene therapy may be potentially administered at a specific localized site by intramuscular injection, skin cell, or synoviocyte replacement, for example. Alternatively, systemic targeting may be achieved by intravenous injection or introduction of engineered bone-marrow cells, for example. Various methods (box) can be used to introduce the genes into any carrier cells (see text for detail).

REFERENCES

Alton, E. W. 1995a. Gene therapy for cystic fibrosis. *Journal of Inherited Metabolic Disease* 18:501-7.

Alton, E. W. 1995b. Towards gene therapy for cystic fibrosis. *Journal of Pharmacy and Pharmacology* 47:351–354.

Annals of the New York Academy of Sciences. Various authors. 1994 ff.

Banerjee, S., Livanos, E., and Vos, J. M. 1995. Therapeutic gene delivery in human B-lymphoblastoid cells by engineered non-transforming infectious Epstein-Barr virus. *Nature Medicine* 1:1303–1308.

Bricca, G. 1995. Sense, antisense, nonsense: Where's the right way? *Journal of Molecular Medicine* 73:417–419.

Cai, D. W., Mukhopadhyay, T, and Roth, J. A. 1995. Suppression of lung cancer cell growth by ribozyme-mediated modification of p53 pre-mRNA. *Cancer Gene Therapy* 2:199–205.

Campbell, T. B., and Sullenger, B. A. 1995. Alternative approaches for the application of ribozymes as gene therapies for retroviral infections. *Advances in Pharmacology* 33:143–178.

Chan, L 1995. Use of somatic gene transfer to study lipoprotein metabolism in experimental animals *in vivo. Current Opinion in Lipidology* 6:335–340.

Chonn, A., and Cullis, P. R. 1995. Recent advances in liposomal drug-delivery systems. *Current Opinions in Biotechnology* 6:698–708.

Colledge, W. H. 1994. Cystic fibrosis gene therapy. *Current Opinion in Genetics and Development* 4:466–471.

Connors, T. A., and Knox, R. J. 1995. Prodrugs in cancer chemotherapy. *Stem Cells* 13:501–511.

Culver, K. W., and Blaese, R. M. 1994. Gene therapy for Cancer. *Trends in Genetics* 10:174–178.

Cunliffe, V., Thatcher, D., and Craig, R. 1995. Innovative approaches to gene therapy. *Current Opinion in Biotechnology* 6:709–713.

Dhawan, J., Rando, T. A., Elson, S. L., Bujard, H., and Blau, H. M. 1995. Tetracycline-regulated gene expression following direct gene transfer into mouse skeletal muscle. *Somatic Cell and Molecular Genetics* 21:233–240.

Dorin, J. R. 1995. Development of mouse models for cystic fibrosis. *Journal of Inherited Metabolic Disease* 18:495–500.

Dunaief, J. L., Kwun, R. C., Bhardwaj, N., Lopez, R., Gouras, P., and Goff, S. P. 1995. Retroviral gene transfer into retinal pigment epithelial cells followed by transplantation into rat retina. *Human Gene Therapy* 6:1225–1229.

Geddes, D. 1995. Designing trials for gene therapy. *Journal of Inherited Metabolic Disease* 18:517–524.

Glorioso, J. C., Bender, M. A., Goins, W. F., Fink, D. J., and DeLuca, N. 1995. HSV as a gene transfer vector for the nervous system. *Molecular Biotechnology* 4:87–99.

Glorioso, J. C., DeLuca, N. A., and Fink, D. J. 1995. Development and application of herpes simplex virus vectors for human gene therapy. *Annual Review of Microbiology* 49:675–710.

Greenberg, D. S. 1995. Gene therapy: Caution and hope [news]. *Lancet* 346:1617.

Herweijar, H., Latendressa, J. S., Williams, P., Zhang, G., Danko, I., Schlesinger, S., and Wolff, J. A. 1995. A plasmid-based self-amplifying sindbis virus vector. *Human Gene Therapy* 6:1161–1167.

Kashani-Sabet, M., and Scanlon, K. J. 1995. Application of ribozymes to cancer gene therapy. *Cancer Gene Therapy* 2:213–223.

Kerr, W. G., and Mule, J. J. 1994. Gene therapy: Current status and future prospects. *Journal of Leukocyte Biology.* 56:210–214.

Kormis, K. K., and Wu, G. Y. 1995. Prospects of therapy of liver diseases with foreign genes. *Seminars in Liver Disease* 1:257–267.

Koths, K. 1995. Recombinant proteins for medical use: the attractions and challenges. *Current Opinion in Biotechnology* 6:681–687.

Kotin, R. M. 1994. Prospects for the use of adeno-associated virus as a vector for human gene therapy. *Human Gene Therapy* 5:793–801.

Lau, C, Soriano, H. E., Ledley, F. D., Finegold, M. J., Wolfe, J. H., Birkenmeier, E. H., and Henning, S. J. 1995. Retroviral gene transfer into the intestinal epithelium. *Human Gene Therapy* 6:1145–1159.

Ledley, F. D. 1995. Nonviral gene therapy: The promise of genes as pharmaceutical products. *Human Gene Therapy* 6:1129–1144.

Martiny-Baron, G., and Marme, D. 1995. VEGF-mediated tumor angiogenesis: A new target for cancer therapy. *Current Opinion in Biotechnology* 6:675–680.

Michel, G., Nowok, K., Beetz, A., Ried, C., Kemeny, L., and Ruzicka, T. 1995. Novel steroid derivative modulates gene expression of cytokines and growth regulators. *Skin Pharmacology* 8:215–220.

Nabel, G.J., Nabel, E. G., Yang, Z. Y., Fox, B. A., Plautz, G. E., Gao, X., Huang, L., Shu S., Gordon, D., and Chang, A. E. 1993. Direct gene transfer with DNA-liposome complexes in melanoma: expression, biological activity, and lack of toxicity in humans. *Proceedings of the National Academy of Sciences (USA)* 90:11307–11311.

Paulus, W., Baur, I., Boyce, F. M., Breakefield, X. O., and Reeves, S. A. 1996. Self-contained, tetracycline-regulated retroviral vector system for gene delivery to mammalian cells. *Journal of Virology* 70:62–67.

Perales, J. C., Ferkol, T., Molas, M., and Hanson, R. W. 1994. An evaluation of receptor-mediated gene transfer using synthetic DNA-ligand complexes. *European Journal of Biochemistry* 226:255–266.

Rios, C. D., Ooboshi, H., Piegors, D., Davidson, B. L., and Heistad, D. D. 1995. Adenovirus-mediated gene transfer to normal and atherosclerotic arteries: A novel approach. *Arteriosclerosis, Thrombosis, and Vascular Biology* 15:2241–2245.

Schadendorf, D., Czarnetzki, B. M., and Wittig, B. 1995. Interleukin-7, interleukin-12, and GM-CSF gene transfer in patients with metastatic melanoma. *Journal of Molecular Medicine* 73:473–477.

Schmidt-Wolf, G., and Schmidt-Wolf, I. G. 1994. Human cancer and gene therapy. *Annals of Hematology* 69:273–279.

Shastry, B. S. 1995. Genetic knockouts in mice: An update. *Experientia* 51: 1028–1039.

Sikora, K., Harris, J., Hurst, H., and Lemoine, N. 1994. Therapeutic strategies using C-erb B-2 promoter-controlled drug activation. *Annals of the New York Academy of Sciences* 71b:115–124.

Tiberghien, P. 1994. Use of suicide genes in gene therapy. *Journal of Leukocyte Biology* 56:203–209.

Wei, Y., Quertermous, T., and Wagner, T. E. 1995. Directed endothelial differentiation of cultured embryonic yolk sac cells *in vivo* provides a novel cell-based system for gene therapy. *Stem Cells* 13:541–547.

Woffendin, C., Yang, Z. Y. Udaykumar, Xu, L., Yang, N. S., Sheehy, M. J., Nabel, G. J. 1994. Nonviral and viral delivery of a human immunodeficiency virus protective gene into primary human T cells. *Proceedings of the National Academy of Sciences (USA)* 91:11581–11585.

Wolfe, J. H. 1994. Recent progress in gene therapy for inherited diseases. *Current Opinion in Pediatrics* 6:213–218.

Yamanaka, R., Tanaka, R., Yoshida, S., Saitoh, T., and Fujita, K. 1995. Growth inhibition of human glioma cells modulated by retrovirus gene transfection with antisense IL-8. *Journal of Neuro-Oncology* 25:59–65.

Yang, J., Tsukamoto, T., Popnikolov, N., Guzman, R. C., Chen, X., Yang, J. H., and Nandi, S. 1995. Adenoviral-mediated gene transfer into primary human and mouse mammary epithelial cells *in vitro* and *in vivo*. *Cancer Letters* 98:9–17.

3

DRUG DISCOVERY BY COMBINATORIAL CHEMISTRY AND MOLECULAR DIVERSITY

3.1. AN OVERVIEW OF THERAPEUTIC TARGETS FOR DRUG DISCOVERY

Most drugs exert their effect either by specifically affecting cytosolic or extracellular enzymes that are vital to biochemical reactions or by binding to membrane-bound protein-based receptors, modifying their function. Since cells communicate with each other and with their milieu through the numerous proteins that stud their surface, cellular receptors are obvious candidates for therapy by molecular intervention. Many of these proteins span and loop in and out of the membrane many times, exposing parts of the same protein to the outside and inside of the cell. These proteins also interact with other proteins at the cell surface. For example, both the T-cell receptor and G-protein coupled receptors are composed of multiple chains that span the membrane, and in both cases they interact with several other protein chains (the CD3 proteins in the case of the T-cell receptor). These proteins are referred to as *receptors* because they are targets through which external signals are transmitted into the cell. The messenger molecule binding to the receptor is called the *ligand.* Insulin and cytokines are examples of ligands that bind to their receptors on the cell surface. To respond to external stimulation or signals, cells need to synthesize novel proteins. For example, activation of

T cells through the T-cell receptors results in the increased production of interleukin-2 (IL-2). In a similar manner a growth factor acting on its receptor will eventually result in the turning on of the genes eliciting cell cycle progression and cell proliferation. The pathways that transmit the external stimuli into the cell and even into the nucleus are called *signal transduction pathways.* These pathways have several ubiquitous intermediates that act as messengers, such as cAMP, inositol phosphates, *G*-proteins, and calcium ions. The binding of ligand to receptor generally results in a conformational change that triggers a cascade of protein-protein interactions, which in turn pass the signal step by step toward the nucleus. The protein intermediates often exist in two forms: phosphorylated and nonphosphorylated. Phosphorylation occurs on Serine, Threonine, or Tyrosine residues and acts functionally as an on/off switch. Some of the signaling intermediates are kinases (enzymes that catalyze the addition of phosphate) and phosphatases (enzymes that catalyze the removal of phosphates). The discovery of low molecular weight compounds that can specifically interfere with these targets and receptors is one of the principal goals of the biopharmaceutical industry. (Examples of the location of different therapeutic targets are illustrated in Fig. 3.1.)

Intracellular Therapeutic Targets

The knowledge that transient compartmentalization of diverse intracellular signaling proteins occurs into large complexes, with separate domains specific for regulatory (binding) and effector (enzymatic) functions, assists greatly in targeting the domain specific for signaling. The rationale for targeting signal transduction pathways has been demonstrated by the analysis of the immunosuppressive drugs cyclosporin A and FK506. Both drugs bind to proteins known as immunophilins within the cell, which in turn interact with calcineurin, a critical component of the T-cell activation pathway that regulates the T-cell specific transcription factor NF-AT. Although the complete process of immunosuppression has yet to be delineated, the rational design of molecules that selectively inhibit calcineurin with potency similar to FK506 is a hotly contested goal for several biotechnology companies.

Another intracellular signaling process is the binding of growth factors to their membrane receptors. These signals activate a cascade of intracellular pathways which in turn regulate fundamental cellular processes such as phospholipid metabolism, arachidonate metabolism, protein phosphorylation, calcium mobilization and transport, and transcriptional regulation. The result is an alteration of cell shape, mobility, adhesiveness to other cells, and DNA synthesis. Aberrations in growth factor-induced events are associated with a number of diseases such as psoriasis and cancer that are due to hy-

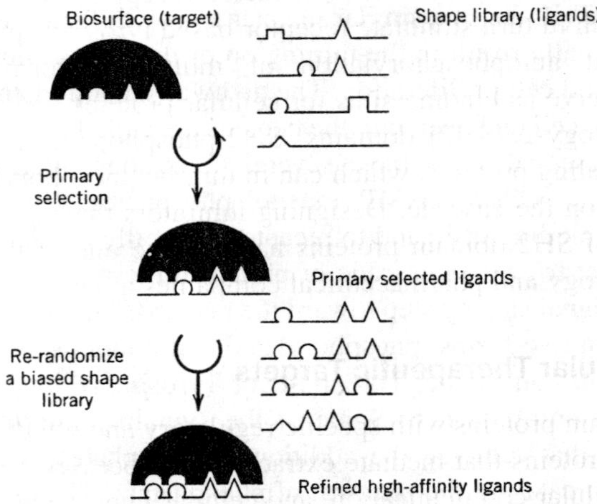

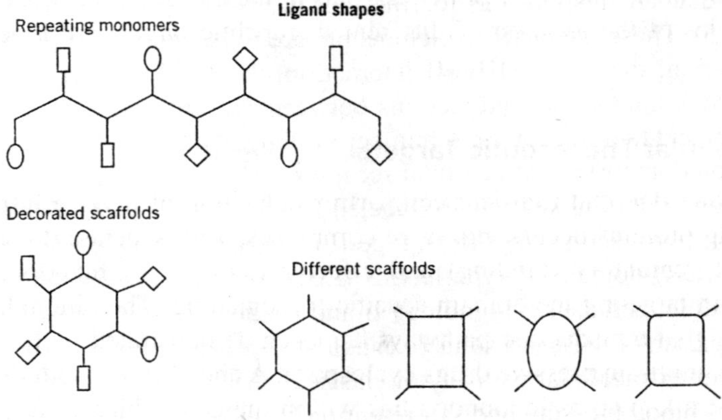

Figure 3.1. Combinatorial chemistry and molecular selection. Combinatorial chemistry is predicated on the ability to isolate targets (the "biosurface" in the figure) and create low-, medium-, or high-throughput assay screens against which large numbers of molecules of differing 3D shape and properties may be exposed (shape libraries). Molecular diversity procedures attempt to generate a large shape library of suitable ligands which are then passed over the biosurface and those that are binders (primary selected ligands) are identified. A systematic or random modification procedure is then used to bias a new library from these ligands. These are again passed over the biosurface and higher-affinity ligands detected. This procedure is repeated until ligands of the requisite affinity (typically in the nanomolar range) are identified. The ligands may be repeating monomers (e.g., peptides, oligonucleotides, or peptoids), scaffolds "decorated" with different functional groups (substituents), or different scaffolds containing different substituents.

perproliferation of cells. Growth factor receptors contain distinct binding sites, which in turn stimulate receptor-based tyrosine kinase activity and the subsequent autophosphorylation of multiple tyrosine residues. These residues serve as binding sites for cellular proteins containing the so-called *src* homology-2 (SH2) domains. SH2-phosphopeptide interactions recruit other signaling proteins, which can in turn be phosphorylated—the "faucet" that turns on the cascade. Designing inhibitors that specifically prevent the function of SH2 domain proteins has become one of the "Holy Grails" of biotechnology and pharmaceutical companies alike.

Extracellular Therapeutic Targets

Multidomain proteins with specific regulatory and effector domains are also found in proteins that mediate extracellular processes, for example, cell adhesion, cellular communication, and cellular interactions that result in complex responses and interactions between T cells and antigen-presenting cells (APCs) that are responsible for the cell-mediated immune response (see Chapter 5). These interactions include, but are not limited to, the CD-2/LFA-3 interaction; the CD-4/MHC II interaction; the integrin alpha/beta receptor/ICAM-1 interaction; interactions between L-selectins, and between the cytokines and their receptors. A number of *domain families* (proteins sharing a common domain) of extracellular receptors have been identified, including the immunoglobulinlike domain (labeled I); epidermal growth factorlike domain (G); fibronectin type I (F1), type II (F2), and type III (F3); a cytokine receptor (CR) domain; C-type lectins (Lec); kringle domains; the complement module; and the *src* homology II and *src* homology III (SH2 and SH3) domains. Each of these is a suitable candidate for interventional drug therapy.

Another extracellular pathway of classical pharmacological interest is the control of blood pressure and salt retention by the renin/angiotensin system. Here the precursor protein angiotensinogen is converted to angiotensin I by the gut enzyme renin, and subsequently to angiotensin II by angiotensin converting enzyme (ACE). Angiotensin II then binds to angiotensin II receptors, leading to a cascade resulting eventually in excessive salt retention and elevated blood pressure. Inhibitors to renin and ACE, and antagonists to the angiotensin II receptors, are all much sought-after molecules of potent pharmacological benefit. Examples of the successful molecular design of ACE inhibitors are discussed in Chapter 4.

Another extracellular target is phospholipase A2. This molecule, isolated from the synovial fluid from patients suffering rheumatoid arthritis (RA), is the rate-limiting enzyme in a cascade resulting in the synthesis of arachidonic acid, a precursor to leukotrienes, prostaglandins, lipoxins, and thromboxanes—all pro-inflammatory mediators. As an extracellular enzyme its selective inhibition is considered essential to mediating diseases such as RA, osteoarthritis, inflammatory bowel disease (IBD), and sepsis.

Yet another target of intense therapeutic interest is the integrin alpha-five/beta-three receptor, also known as the *vitronectin receptor* (VNR). High levels of VNR expression has been implicated in the proliferation of vascular endothelial cells in solid tumor tissues (Brooks et al. 1994). Excessive vascularization (also known as angiogenesis) is a hallmark of a number of disease states, including solid tumor cancers, macular degeneration, and diabetic retinopathy. The mechanism of angiogenesis requires the activation of quiescent endothelial cells by secreted growth factors such as basic fibroblast growth factor (bFGF), platelet-derived growth factor (PDGF), vascular endothelial growth factor (VEGF), and interleukin-8 (IL-8). In addition migration of endothelial cells requires their adhesion to extracellular matrix proteins such as vitronectin, fibronectin, laminin, and collagen. Cellular adhesion is dependent on a family of cell adhesion molecules (CAMs) named the *integrins*. Cell surface adhesion receptors play crucial roles in blood clotting, wound healing, and inflammation. VNR makes a particularly attractive target for inhibiting angiogenesis, since VNR is only expressed in proliferating vascular endothelial cells and not in other cells of the body. Antagonists to VNR appear to prevent angiogenesis and inhibit tumor growth and metastasis in animal models by inducing programmed death of primary endothelial cells.

3.2. COMBINATORIAL CHEMISTRY IN DRUG DEVELOPMENT

The history of pharmaceutical drug development to the types of targets described above has traditionally focused on the selection of a disease for which a well-defined and reasonably large market exists. Typically little was known about the biology of the disease at the molecular or cellular level (prior to the advent of genetic engineering and other technological advances). Most drugs developed prior to the biotechnology revolution were developed through a combination of luck and intuition regarding the basis of drug function and optimization, coupled with systematic synthesis and screening procedures of biologists and chemists. (Optimization here refers to the increase in affinity of the drug to its receptor together with the elimination or minimization of any toxic side effects.) Most drugs discovered in the past originated from the screening of samples from the soil and elsewhere (where fungi and bacteria often secrete potent pharmacologics), followed by isolation, chemical synthesis, and elimination of side effects. Others came from screening large numbers of isolated compounds from a variety of nat-

PHARMACOLOGICAL DEFINITIONS OF RECEPTOR-EFFECTOR INTERACTIONS

An *agonist* is a substance that interacts with a specific cellular constituent (the receptor) and elicits an observable response. An agonist may be an endogenous molecule such as a neurotransmitter or hormone, or it may be a synthetic drug.

A *partial agonist* typically acts on the same receptor as other agonists in a group of ligands (binding molecules) or drugs. However, regardless of its dose, it cannot produce the same maximum biological response as a full agonist.

The *intrinsic activity* of an agonist is defined as the proportionality constant of the ability of the agonist to activate the receptor compared to the maximally active compound in a series under analysis. Thus the intrinsic activity is unity for a full agonist and zero for an antagonist (see below). The intrinsic activity is comparable to the Michaelis constant (K_m) of enzymes.

An *antagonist* inhibits the effect of an agonist but has no biological activity of its own in a given biological system. It may compete for the receptor site or act on an allosteric site, in which case antagonist binding "distorts" the receptor (typically by causing a conformational change), preventing the agonist from binding to it.

The *median effective dose* (ED_{50}) is the amount of drug required for half-maximal effect, or producing an effect in 50% of a group of experimental animals. It is usually expressed as mg/kg body weight.

The *in vitro* ED_{50} is expressed as a molar concentration (EC_{50}) rather than an absolute amount.

The *median inhibitory concentration* (IC_{50}) is the concentration at which an antagonist exerts its half-maximal effect.

ural sources such as plants (commonly referred to as *natural products screening*), while still others were from the rigors of synthetic chemists. As a result large, established pharmaceutical companies have dominated the industry thanks to their accumulation, over time of proprietary databases of compounds from which they can find "leads" against newly targeted diseases. These compilations have mostly unknown function, but the hope is that a few compounds will be found to be of potent pharmacological benefit against a novel target. Even if such a compound is found, the process of optimization requires medicinal and synthetic chemists to deduce which *sub-*

stituents (functional groups of atoms) in candidate lead molecules can be modified to retain potency yet reduce or eliminate side effects such as toxicity and crossreactivity. This procedure required the time-consuming sequential synthesis of each identifiable candidate (lead) molecule. In practice, this means altering the lead compound by systematically trying different *functional groups* (arrangements of reactive atoms; e.g., bond-forming carbonyls, amides, or hydrogen-binding nitrogen or oxygen atoms) in order to improve potency. The pharmacological activity of a drug is usually defined as its ability to inhibit, agonize, or antagonize a target (receptor) by binding to it with requisite affinity (usually considered to be in the nanomolar to subnanomolar range). Binding is achieved by a stereoelectronic interaction whereby the receptor recognizes the three-dimensional arrangement of functional groups and their electron and charge density. The collection of the relevant groups and their three-dimensional arrangement is known as the *pharmacophore* (see the Glossary at the end of the book and the following chapter); the complementary groups in the receptor are collectively known as the *receptor map*.

In contrast to the pharmaceuticals, biotechnology companies have used the full power of genetic engineering technology to develop an understanding of a disease state at the molecular level. Mapping the molecular pathology of a disease by genetic analysis allows the identification of the molecules in a cellular cascade that results in a diseased cell or tissue. Targets might include genes, gene regulatory factors, gene products (including enzymes and regulatory or signaling proteins), and both intracellular and extracellular receptors. The ability to clone, express, and isolate virtually any desired target has led to an explosion in the number of therapeutic strategies possible—and a concomitant explosion in the number of biotechnology companies bent on discovering molecules that moderate the effects of target proteins. Biotechnology companies have also developed rapid *in vitro* methods for generating, screening, and amplifying drug candidate molecules using the *same* basic building blocks of life: nucleic acids, amino acids, and small organic molecules. The goal of this technology is to develop drug molecules from common chemical building blocks that explore the three-dimensional "shape space" of their targets and bind to them with the requisite affinity and specificity. (At the molecular level this requires finding significant steric and electronic complementarity between the target and the drug.) The general name for a combination of related strategies allowing the rapid construction of large, chemically diverse molecule libraries is *molecular diversity* (MD) (Stone 1993). MD involves the rapid synthesis and screening of large compound libraries consisting of arrays of novel structures made by the random or directed synthesis of a combination of smaller molecular building blocks. MD relies on two technological developments that have been key to the competitive advantage of the biotechnology industry: the ability to isolate target molecules in pure, crude extract or whole cell *in vitro* assay screens, and the development of robotics and instrumentation to perform high-capacity screening on microtiter plates in a rapid and automated fashion. Molecular diversity originated with attempts at mimicking the natural diversity of the immune system to create antibodies and T-cell receptor of almost limitless variety; this concept has now been extended to develop catalytic antibodies that can catalyze reactions of important synthetic value to medicinal chemists (Schultz and Lerner 1995).

Today pharmaceutical companies themselves use MD as an extension of traditional library screening and medicinal chemistry, while biotechnology companies have used MD techniques to progress from the molecular biology of "large" molecules (typically proteins or DNA) to "small" molecules (peptides, DNA fragments, carbohydrates, etc.) The promise of MD is in the fact that emerging companies can leverage proprietary methods of creating libraries using novel chemical classes and thus compete against large established pharmaceuticals. MD shows promise in the discovery of novel antibodies, proteins, rizbozymes (autocatalytic RNA molecules), oligonucleotides, peptides, carbohydrates, and small organic molecules (representative examples of the application of MD to each of these is described in more detail below). MD has become a compelling procedure that promises to provide a quantum leap in improvement in the efficiency with which companies seeking new drug compounds can generate prioritized leads.

The basic strategy of MD involves the synthesis of large compound libraries from peptides, oligonucleotides, carbohydrates, to synthetic organic molecules. The choice of synthetic methods may include biological, enzymatic, or chemical approaches. Critical parameters in developing a MD strategy include subunit diversity, molecular size, and library diversity. Using amino acids, subunit diversity is limited to 20; thus a linear peptide library of N-mers will have 20^N compounds. (A library of 4-mers can have 20^4 or 160,000 compounds.) Building a larger library requires longer peptides, nonstandard amino acids, or a different combination of synthesis methods. Nucleotide-based libraries offer fewer subunits, but the ability of these compounds to adopt a great number of 3D structures and act as templates for their own synthesis permits directed molecular evolution (DME; see below) to be used to select candidate drugs. Carbohydrates and organic molecules can "pack" greater subunit diversity because they are not restricted to linearity (most carbohydrates are branched-chain molecules since they permit linkages at several sites around the sugar ring), but these types of libraries are often the hardest to synthesize and screen. Small organic molecules hold perhaps the greatest benefit for therapeutic MD, since history has shown that these molecules make the best drugs (i.e., they are small, easily ingested, stable, most likely to find their target receptor, and usually present the fewest side effects). However, they are perhaps the most challenging from which to create large combinatorial libraries, since they require special chemistries to ensure synthesis of a structurally diverse chemical library.

The field of molecular diversity has generated its own language to describe the properties of the libraries generated. These include *shape* (the 3D features of a molecule that govern its interactions with other molecules), *shape library* (a population of potential ligands generated from randomized biopolymers), *shape space* (all of the shapes that can be assumed by a given sequence, or population of sequences), *target* (molecule used to select a ligand from a shape library), *mimetic* (ligand that resembles the shape of another molecule and competes with it for binding), and *degeneracy* (maximum number of potential ligands present in a given shape library). The goal of MD methods is to rapidly search shape space for mimetics that bind to the target with the requisite affinity and then to optimize these lead molecules by either random or directed design strategies. These concepts are illustrated in Fig. 3.1.

One form of MD is directed molecular evolution, or DME, which can be applied to molecules that code for their own synthesis, for example, self-replicating molecules such as oligonucleotides. DME makes no attempt to determine *a priori* what functional groups need to be modified to a molecular scaffold in order to find lead molecules; instead the procedure simply selects the "fittest" molecules in a given pool. In a typical DME experiment a self-replicating molecule is subjected to random mutagenesis across some portion of its length. The molecules are then screened against a target, and those that bind are amplified, using the sequence of the mutagenized molecule as the template for amplification. These molecules are then released from the immobilized target and used in another round of mutagenesis and amplification, where the mutagenesis is restricted to a neighboring region in the structure. DME uses the fact that successive rounds of screening and amplification of binders *in vitro* accurately mimics evolution, except that numerous rounds of evolution are possible in very short cycles. DME has been applied to antibody CDR loops expressed by phage display (Smith 1985; Parmley and Smith 1988; Scott and Smith 1990), to autocatalytic RNA molecules (ribozymes) that have been evolved for novel reactions (Szostack, 1993), and to oligonucleotides that can be amplified by PCR methods. An outline of the DME strategy is illustrated in Fig. 3.2.

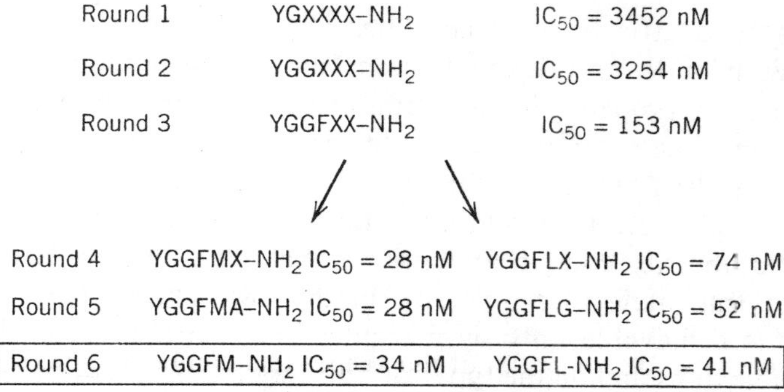

Figure 3.2. Directed molecular evolution: Evolution of nanomolar affinity substrates. Using the Houghten SPCL procedure a series of peptides are randomly generated and screened for their ability to competitively inhibit against a known encaphalin-binding peptide, known as DAMGO, that binds the opiod μ receptor. Starting with a micromolar lead (YGXXXX-NH2), six rounds of directed molecular evolution resulted in two pentameric peptides with affinities in the range 30–40 nM.

3.3. METHODS FOR CREATING MOLECULAR DIVERSITY

In general, all molecular diversity approaches utilize sequential application of combinatorial selection to obtain high-affinity ligands against a desired target. In the first round a randomized "shape library" generated in one of a multitude of ways is screened against a target to select ligands that bind to one aspect of the target biosurface. The next round uses a suitably biased shape library containing the same primary ligand binding sites but locally randomized to select for secondary site ligands. Through an iterative process of selection, screening, and localized (directed) randomization, it is possible to generate high-affinity ligands against multiple independent surfaces on the target molecule. Early methods of molecular diversity were developed for oligonucleotide or peptide libraries, where initial leads could easily be modified ("mutated") to serve as templates for the next round of optimized synthesis. Because of the vast range of shapes that oligonucleotides can adopt and their ease of synthesis (oligonucleotides and peptides can now be manufactured in fully automated synthesizers), some companies such as Affymax and Gilead Sciences are continuing this approach. Oligonucleotides and oligopeptides can adopt a huge variety of "shapes" (conformations) allowing them to juxtapose the functional groups that can recognize and bind to a target by forming the requisite bonds and interactions. Large libraries of these molecules allows isolation of molecules that have a high degree of specificity and affinity to a particular target. However, leads discovered from peptide or oligonucleotide libraries still require extensive modification to produce suitable drug candidates since biopolymers are typically unsuitable for use as stable, orally active drugs. Further the available building blocks are limited, even allowing for unnatural enantiomers or chemically modified amino acids or nucleotides. Also oligomeric libraries contain a repetitive backbone linkage, reducing the degree of true molecular diversity in their structures. Real chemical diversity is achievable only by removing the restrictions on the bond-forming reactions and on the types of building blocks that can be used. If this can be achieved, the resulting monomer libraries would have the same advantages and uses as oligomer-based libraries with the added benefit that the target structure would be limited only by the creativity of the synthetic chemist designing the bond-forming chemistry. Core criteria for successful library design include a stable population of low molecular weight chemical entities that are free of reactive and toxicity-causing functionality. Libraries may be general purpose or focused about a pharmacophore (a pharmacologically important spatial distribution of atoms and functional groups, Sepetov et al. 1995; pharmacophores are described in more detail below) in order to detect more potent analogs or expanded to find new pharmacophores for novel targets obtained by molecular cloning and gene expression. (Examples of pharmacophores of medicinal interest include the benzodiazepenes, beta-lactams, imidazoles, phenethylamines, and others. Some of these are illustrated in Fig. 3.3.)

MD and combinatorial chemistry also require technology for the efficient organization and storage of the large volumes of screening data available from the robotics assays. Essential parameters for a company undertaking a MD effort include the choice of targets to be screened and their availability, the type of building blocks to use in the combinatorial synthesis, the initial choice of combinatorial libraries to synthesize (libraries should be diverse but not so diverse that any given molecule is titered at too low a concentration to be detected), whether or not the molecules are "tagged" (individually labeled), the optimization procedure (whether molecules are directly optimized by user-selected criteria or randomly optimized by directed molecular evolution), and the methods for data collection, storage, and analysis. We will consider examples of each of these below with reference to particular companies that are involved in their development.

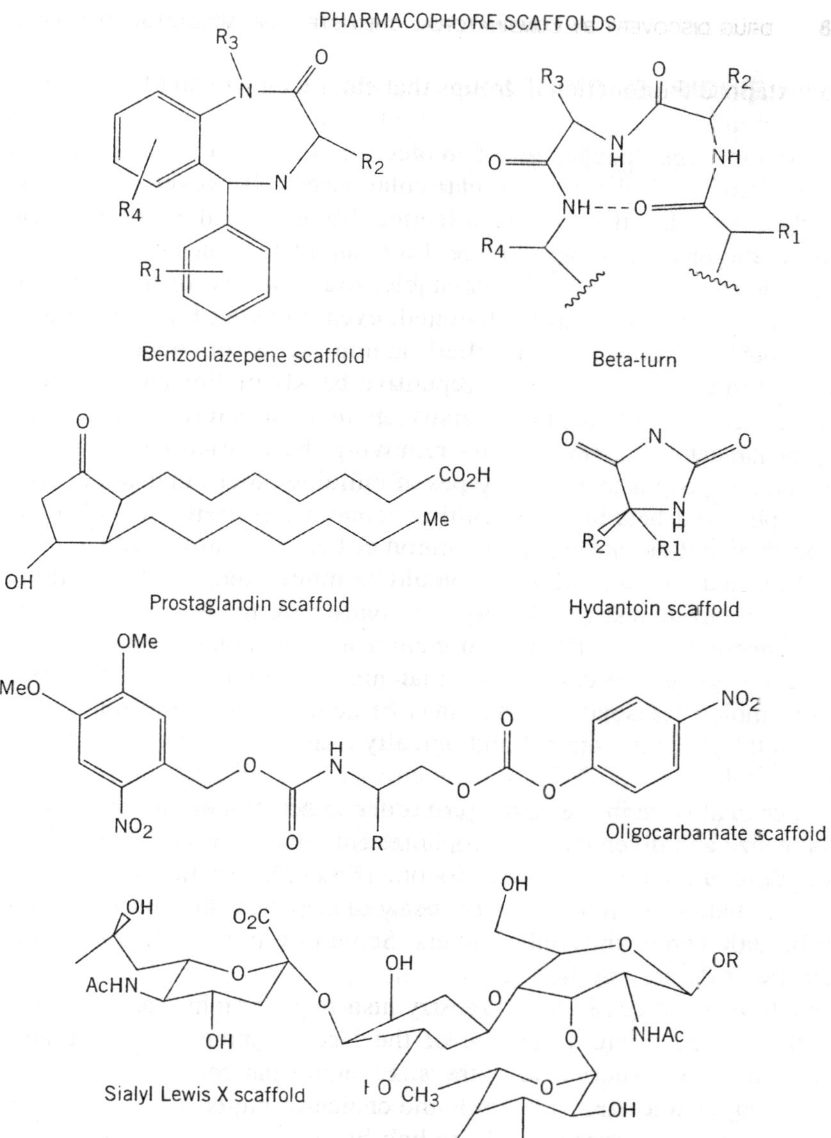

Figure 3.3. Pharmacophore scaffolds. Some pharmacologically relevant scaffolds illustrated in the figure that are now amenable to solid-phase synthesis include the benzodiazepenes (pharmacologically these can be used as anxiolytics, sedatives, cholecystokinin A,B antagonists, opiod receptor antagonists, HIV *tat* antagonist, PAF antagonist); the prostaglandins (used as antiulcer, bronchodilators, abortafacients, antithrombotics); beta-turn blockers (used in receptor/peptide hormones, antibody CDR turns, enzyme/substrate recognition); alpha- and beta-blockers (channel phosphorylation or glycosylation); and carbohydrates such as Sialy Lewis X (useful as antitumor agents, cardiac glycosides, carbohydrate receptors).

3.4. PRINCIPLES OF COMBINATORIAL CHEMISTRY

One of the biggest advantages of combinatorial chemistry over classical synthetic chemistry is that is can lead to compounds that otherwise might not be synthesized using the traditional methods of medicinal chemistry. For instance, the benzodiazapenes possess a range of biological activities, interacting with a broad range of biological receptors in addition to the diazapene receptor (which is responsible for the tranquilizing and antianxiety effects of the Roche inhibitor Valium). Microtiter plate instrumentation has been adopted to the simultaneous solid-phase synthesis and screening of the benzodiazapenes (and hydantoins), allowing large libraries to be screened. The establishment of a combinatorial synthesis chemistry has allowed libraries of related compounds to be readily created, such as the prostaglandin family or the alpha- and beta-blockers. In the solid-phase benzodiazapene synthesis procedure, a 2-aminobenzophone derivative is linked to a solid support and then reacted with an FMOC (fluoroenylmethoxycarbonyl)-protected amino acid. A cyclization step followed by cleavage from the support with an alkylating agent releases a 1,4-benzodiazepene derivative that can be isolated and screened against the desired target (Bunin and Ellman 1994). Biologically active molecules isolated in this way include inhibitors to the opiod receptor and to the HIV *tat* protein. Solid-phase synthesis of hydantoins is illustrated in Fig. 3.4.

A critical issue in the development of combinatorial libraries is the requisite size and diversity of a small molecule library to achieve a thorough sampling of the biological activity of a class of molecules against a given target. Spatial separations can typically be achieved for libraries of 10,000 to 100,000 compounds only, whereas pooling strategies will be needed for libraries that are any larger. To create these larger libraries, compounds have to be synthesized together, and the activity of individual compounds in the pooled mixture deconvoluted by competitive binding assays or other methods. Alternatively, the compounds can be labeled while they are being synthesized (tagging) so that their identity can be determined if they have interesting biological properties. Still and coworkers (Columbia University) have developed a chromatographic "bar code" that unambiguously identifies every compound in a library. Using "split synthesis" method (see sidebar) in tagging the molecules being synthesized, chemists attach to the beads both the step number and the chemical reagent used in that step. The tagging molecules themselves are not sequentially connected to each other. With 20 such tags, 2^{30} molecules can be uniquely encoded. The tags are usually structurally related molecules that an be analyzed by a variety of techniques such as electron capture capillary gas chromatography. Chemically tagged combinatorial

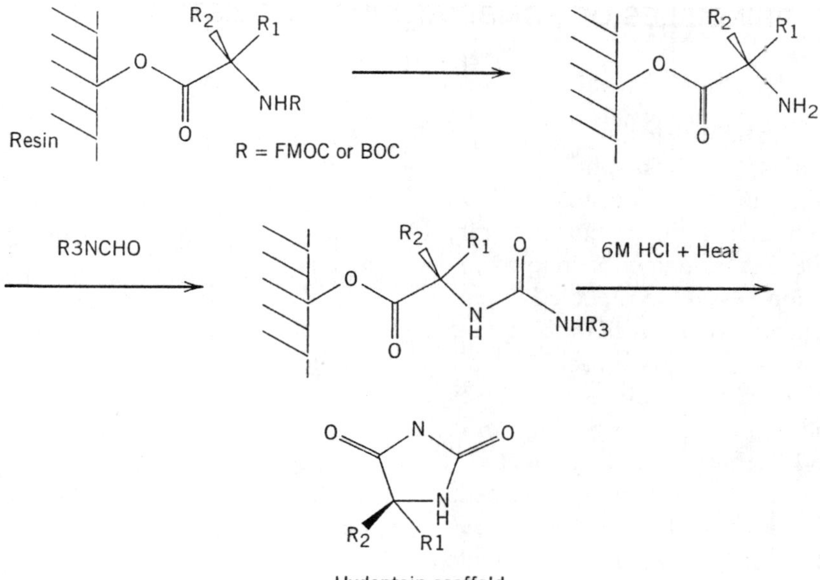

Resin

R = FMOC or BOC

R3NCHO

6M HCl + Heat

Hydantoin scaffold

Figure 3.4. Solid-phase synthesis of Hydantoins. Hydantoins are an important class of therapeutically relevant compounds and their solid-phase synthesis by the "diversomer" method now allows exploration of combinatorial properties of different R-groups. The procedure can also be applied to other pharmacologically relevant molecules such as the benzodiazepenes (see Fig. 3.5).

libraries of compounds are now being made available by companies such as Pharmacopaea (Burbaum et al. 1995). More recently the use of radio-emitting microchips to tag compounds during synthesis has been implemented (Ontogen Corp. and Irori Corp.). The advantage of these is that they do not need to be synthesized along with the compounds themselves and that they are chemically inert, enabling a harsher set of reagents to be used during synthesis. (In the Ontogen approach a radio scanner registers both the identity of a capsule and the contents of each reagent beaker it enters.) The concept of tagging is illustrated in Fig. 3.5 (see page 84).

The identification of compounds in a combinatorial chemical library is typically by one of three methods:

1. In the case of peptides or oligonucleotides, it is the sequence of the molecules themselves. (The oligonucleotides can easily be amplified and sequenced using PCR techniques.)

SYNTHESIS STRATEGIES FOR COMBINATORIAL CHEMISTRY

Two synthesis technologies allow the rapid creation of combinatorial libraries: "mix and split" synthesis and "parallel" synthesis. In the former, microscopic resin beads are divided into *N* reaction vessels, one for each monomer in the monomer library. (*N* would be 20 in peptide synthesis.) The first monomer is coupled to the resin, and then the beads in each reaction vessel are repooled. The cycle of divide, couple, and repool is repeated several times (*n* times to create an *n*-mer library), using a suitable deprotecting group each time a new monomer is added. Separation into the *N* reaction vessels guarantees an equimolar representation of each of the *N*-mers at each position in the combinatorial library (Schnorrenberg and Gerhart 1989; Geyson et al. 1984).

Once the libraries are created, active molecules can be identified by a number of techniques, one of the most common being the use of a monoclonal antibody (MAb) or soluble receptor. A second (idiotypic) antibody or enzyme conjugated to a fluorescein conjugate is then targeted against the binding Mab, allowing its ready identification in the light microscope (this is the basis of the classical ELISA test). The stained beads are selected and their attached molecules sequenced (if they are peptides) using an automated peptide sequencer. In the case where the library is being made from nonsequenceable organic compounds, the beads must be "labeled" or "tagged"—usually with a linear peptide or nucleotide sequence whose order maps to the order of synthesis of the monomers in the library (Nielsen et al. 1993–see text for details).

The solid-phase synthesis method is especially suitable for peptide and oligonucleotide libraries. However, this method requires the simultaneous parallel synthesis of all possible *N*-mer combinations: To create a hexamer library from *N*-mers where $N = 20$ would require 20^6 reaction vessels. Instead, randomization is carried out at the first 2 positions (say); requiring $20^2 = 400$ reaction vessels, then the remaining 4 positions are filled randomly by pooling all 20 monomers in equimolar amounts. This creates 400 mixtures with 2 random monomers at the same positions in each. The library can be screened by pooling these mixtures into 96-well microtiter plates, using an ELISA-based (or other) assay to detect positive wells. The mixtures in a given well must then be deconvoluted (Erb et al. 1994) to identify the tight-binding molecule, such as by a competitive binding assay that distinguishes between molecules in a given mixture.

"Hits" from cycle 1 of this synthesis can be improved by systematically randomizing positions 3 and 4 in the hexamer into 400 reaction vessels and repeating the process. The cycle continues until all positions in the hexamer library have been randomized. Critical to the success of this method is the sensitivity of the bioassay used to define the best binder at the two defined positions. In a library of 20^5 pentamers the concentration of one specific monomer in a 1 mg/ml solution is only 0.5 nM. Separating a lead from the background in this case is only possible under very stringent conditions, leading many to attempt methods of creating smaller, more molecularly diverse sample libraries that can be used in screening. Once initial binders are detected, more focused libraries containing molecular properties similar to the lead can be generated for subsequent passes. Methods of measuring diversity are described below.

In the case of small molecule parallel synthesis, compounds are synthesized in separate vessels in liquid-phase without remixing, typically by using a robotic arm to add reagents to different wells of a 96-well microtiter plate. Hits from the library are then identified by their position on the well plate. Multiple parallel synthesis restricts the production of compounds to smaller numbers then "split" or "mixture" synthesis. However, greater quantities of the compounds are created, and there is no need to deconvolute or tag the individual compounds. Also, since they are plated as single compounds per well, activities measured from high-throughput screening can be correlated directly to the individual compounds for subsequent structure activity relationship (SAR) studies.

2. In the case of chemical compounds, they may be tagged by peptides or oligos, by photoreactive, radio-emitting, or optically active tags.
3. In either case they may be tagged by spatial location, using polyethylene or other pins arrayed on microtiter plates, which are then exposed to fluorescently tagged drug targets. (In the case of the Affymax VLSIPS process described below, the arrays are custom-designed photolithographic chips.)

3.5. MOLECULAR DIVERSITY OF PROTEINS

Phage Display of Antibody F(ab) Fragments (Scripps Research Inst.; Ixsys, Inc.)

Molecular diversity as an industrial technology can trace its origins from attempts to express the molecular diversity of the mammalian immune system in nonmammalian cells (McCafferty et al. 1990; Kang et al. 1991). This initially resulted in a vector system that produced F(ab) fragments of mam-

SPLIT SYNTHESIS METHOD (SELECTIDE PROCESS)

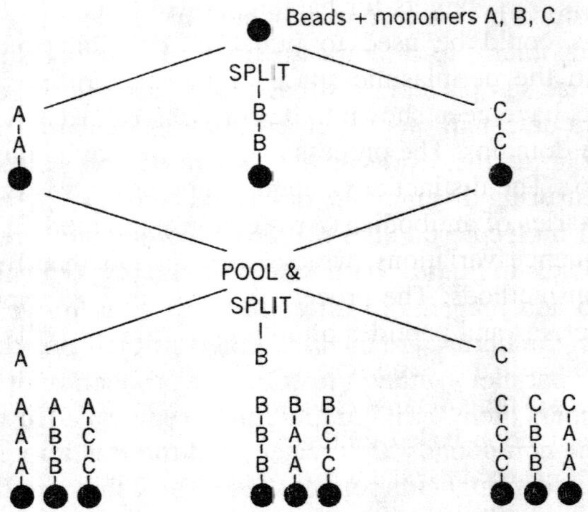

ENCODED LIBRARY SYNTHESIS ("TAGGED" SYNTHESIS)

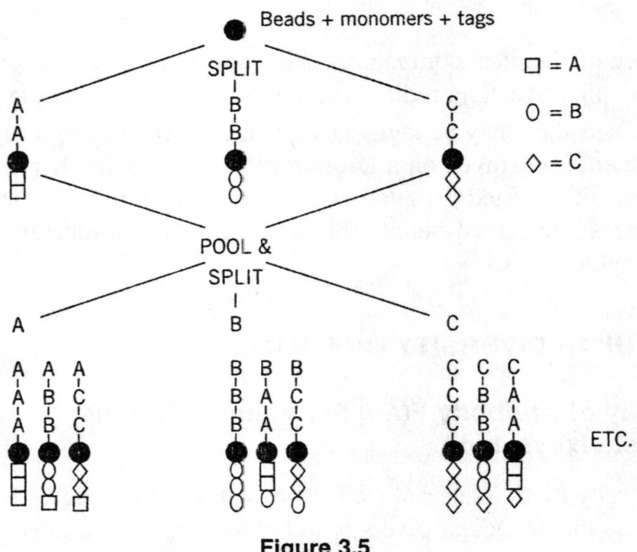

Figure 3.5

malian antibodies displayed as a VHCH1-pVIII fusion protein on the surface of M13 bacteriophage, or as a gene III-VCHC1 fusion protein in phage Fd (pVIII and the gene III protein are capsid proteins that are part of the exposed phage shell). These bacteriophage, when randomized, generate literally millions of antibody F(ab) fragments, and when the resultant cultures are immobilized they can be used as targets for hapten/antigen selection. In addition the phage vectors could be used to infect *E. coli* and free F(ab) fragments secreted into the periplasmic space of the bacterium. The expressed F(ab) fragments have been shown to be correctly folded and assemble as complete protein domains. The process of phage display is illustrated in Fig. 3.6 (see page 86). The distinct advantage of this process is that very large combinatorial libraries of antibodies (Abs) can be expressed, of the order of 10^5–10^8 Ab sequence variations, which is far greater than by traditional hybridoma fusion methods. The process is not limited to expressing antibodies however, proteins and peptides of all types are now routinely being randomized by "phage display" and screened against therapeutic targets. These phage expression systems lend themselves to efficient protein engineering by mutagenesis. Codon-based mutagenesis is also possible by randomly mutating the genes three bases at a time (ensuring that each mutation substitutes a unique amino acid). In this way mutagenesis, screening, and amplification of proteins against ligands (or of antibodies against haptens and antigens) can be accomplished extremely rapidly. Note that this procedure is a dramatically different one for engineering changes into proteins from the one described in the next chapter (the protein engineering by *site-directed* mutagenesis described there presupposes a detailed knowledge of the protein structure before a change can be engineered, whereas random codon-based mutagenesis requires no prior knowledge of the protein but simply creates all possible changes and then selects the best according to some assay). An issue unique to expressing combinatorial libraries of anti-

←

Figure 3.5. Split synthesis method (Selectide process). In the Selectide process, monomers are attached to beads, and these molecules are pooled and then split into separate wells. In the second step, additional monomer of one type only is added to each of the wells, the attachment chemistry performed, and the wells once again pooled and split. The procedure of "pool and split" is iterated a number of times until a representative sampling of all possible monomer attachments have been synthesized. In the "tagged" approach, a molecular "tag" or "label" (e.g., an amino acid or nucleotide whose sequence can readily be determined—*A, B, C* in the figure) is also attached to each bead at the same step as the attachment of the monomer. Deconvolution of the compounds in each well at the end of the procedure is then simply one of sequencing this "molecular bar code" in order to determine the particular monomers identified in an active molecule.

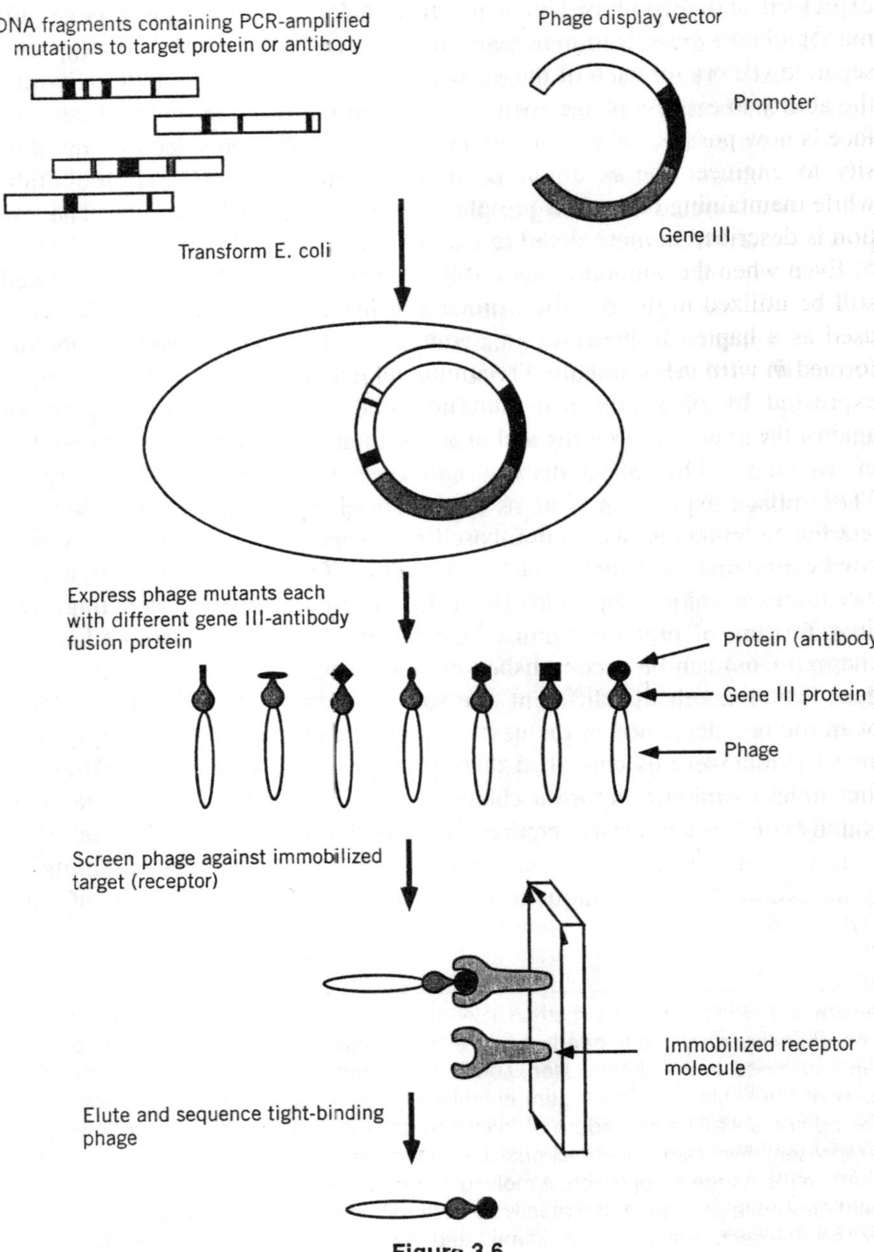

DNA fragments containing PCR-amplified mutations to target protein or antibody
Phage display vector
Promoter
Gene III
Transform E. coli
Express phage mutants each with different gene III-antibody fusion protein
Protein (antibody
Gene III protein
Phage
Screen phage against immobilized target (receptor)
Immobilized receptor molecule
Elute and sequence tight-binding phage

Figure 3.6

body genes in phage is that both the H (heavy) and L (light) chains must be expressed and recombined in a manner analogous to the process of immunoglobulin gene rearrangements in mammalian cells. By creating two separate vectors for each of the gene types and recombining them in phage, the ability to express phage with recombinant F(ab) fragments on their surface is now possible. For instance, Ixsys, Inc. has used condon-based diversity to engineer the randomized humanization of therapeutic antibodies while maintaining or even improving their specificity. (Antibody humanization is described in more detail in the next chapter.)

Even when the antibodies are not themselves used as new drugs, they can still be utilized in the selection process. Initially a compound of interest is used as a hapten to generate an immune response. This step may be performed *in vitro* using modified recombinant antibody libraries such as those expressed by phage display. Combinatorial selection is then performed against the generated monoclonal or polyclonal antibodies. The combination of hapten and antibody generates a "hapten epitope tag" on the antibody CDR surface that can be used against a RNA (or other) shape library to select for molecules that compete with the original hapten. Mimetics with desired activities could be subject to structural analysis, and their tertiary struture used as the basis to model small chemical compounds with the desired drug properties. (This technique is discussed in more detail in the next chaper.) Companies such as Enzon and Creative Biomolecules have used gene fusion methods and PCR to covalently attach the light and heavy chains of antibody F(ab) fragments with a repeating "linker" peptide in order to create so-called single-chain variable region fragments (scF$_V$'s) that may be either monovalent (engineered to bind to one antigen or hapten) or bivalent (capable of binding two different antigens or haptens). These molecules may

Figure 3.6. Phage display of proteins and antibodies. Phage display technology uses genes containing a spectrum of mutations directed to a target gene product (protein, antibody, or peptide) that are incorporated into a phage display vector containing a promoter and the gene III capsid protein of phage. This construct is used to transform *E. coli* cells (the phage host) resulting in a huge amplification of phage particles (typically 10^{10} or more). The phage express the gene III and target proteins as a fusion protein on its surface, where the protein is available for binding provided that the phage particle has been immobilized. The phage are then passed across a screen containing immobilized target (receptor) molecules, and those that bind are eluted and sequenced. Molecular evolution is performed by iterating successive rounds of amplification, screening and selection, resulting in phage expressing mutants that bind their target receptor with nanomolar affinity or better. The process can be made more efficient by restricting the mutations suffered by the phage to codon-based mutations (i.e., every mutation results in the substitution of a sense codon).

serve as diagnostics in their own right or be coupled to a toxin or enzyme in order to neutralize pathologically diseased cells such as cancer cells.

Directed Evolution of Proteins of Therapeutic Significance (Protein Engineering Corp., Cambridge, MA)

The phage display technology pioneered for the generation of antibody diversity has been used to raise single-chain antibodies (antibodies whose light and heavy chains are covalently attached to one another by a peptide linker) and nonantibody proteins such as human growth hormone (Bass et al. 1990). Each phage particle links a particular gene carried with the particle to a gene-encoded protein displayed on the phage surface, typically as a fusion protein with the phage's own capsid (shell) proteins (de la Cruz et al. 1988). Directed evolution of these proteins can be accelerated greatly by constructing a library of phage displaying a multitude of variants of the target protein and fractionating (selecting) the variants that display proteins with the highest affinity to a desired target (Lowman et al. 1991). Since phage titers of up to 10^{13} plaque-forming units(pfu)/ml can be obtained readily, a vast number of phage can be examined. Also fusion phage can be further amplified and subjected to additional rounds of directed evolution and selection (Markland et al. 1991). Phage display is illustrated in Figure 3.6 and the directed molecular evolution of peptide sequences is illustrated in Fig. 3.7.

Using this technology, inhibitors to human neutrophil elastase (HNE) have been "engineered" by designing and producing a library of phage-displayed protease inhibitory domains derived from wild-type bovine pancreatic trypsin inhibitor (BPTI), followed by fractionation of the library for binding to the target elastase by directed molecular evolution (Roberts et al. 1992). HNE was chosen because it is an abundant serine protease involved in the elimination of pathogens and connective tissue restructuring. Extensive destruction of lung tissue results from uncontrolled elastolytic activity of HNE, when it is not itself inhibited by alpha-1 antitrypsin inhibitor. This may occur either due to hereditary reasons or by oxidation of the inhibitor (the cause of smoker's emphysema). Using this technology, Protein Engineering Corp. scientists have been able to detect variants to HNE selected via DME in the picomolar range—10 million times higher than the parental protein and 50× higher than the best reported reversible inhibitor to this protease. Combinatorial proteins were obtained by constructing a gene III fusion phage comprising 1000 displayed engineered protease inhibitors based on wild-type bovine pancreatic trypsin inhibitor (BPTI). BPTI was chosen because it is a small, easily expressed protease inhibitor for which the 3D structure is known. BPTI has also been shown to be safe when administered as a drug. A "variegated" oligo was designed to mutate the residues

Length	Peptide	Number of combinations
2	AC–OO–NH_2	400
3	AC–OOO–NH_2	8,000
4	AC–OOOO–NH_2	160,000
5	AC–OOOOO–NH_2	3,200,000
6	AC–OOOOOO–NH_2	64,000,000
7	AC–OOOOOOO–NH_2	1,280,000,000
8	AC–OOOOOOOO–NH_2	25,600,000,000

Possible SPCLs built using hexamers

AC–OOOOOO–NH_2 (400 groups) 64,000,000 individual hexamers

1 AC–OOXXXX–NH_2
1 AC–XXOOXX–NH_2 160,000 HEXAMERS PER POOL
1 AC–XXXXOO–NH_2

Ac— Acetylated carboxy-terminus
O— 20 individual L-amino acids
X— Combination (mixture) of 20 individual L-amino acids
NH_2— Amino terminus

Figure 3.7. Combinatorial peptide libraries. With a shape library restricted to the 20 naturally occurring amino acids, there are still 2.56×10^{10} different peptide octamers possible by substituting for every possible amino acid at each of the eight positions. Even with the ease of peptide synthesis and phage display methods, it is still impossible to screen every one of these. However, using liquid-phase pooling strategies, it is possible to generate hexamers where at particular positions every possible amino acid is substituted, while at other positions some combination of amino acids are selected from a pool of all possible substitutions. In the figure, O's represent every one of the 20 possible amino acids, while X's represent a randomly selected amino acid. With the Houghten procedure this results in 160,000 hexamers per pool, a number more readily screenable by current high-throughput screening procedures.

15–19 of wild-type PBTI, as these are known to be the principal effectors of its inhibitory effect. Cassette mutagenesis was used to transform phage plaques exhibiting about 90% of the potential inhibitors for this region (~900 of 1000 BPTI variants). The highest affinity mutant has a 50-fold higher affinity for HEN than that of antileukoprotinase, the best characterized inhibitor to HNE.

Protein Engineering Corp. believe this technology can be used in a general manner to improve the affinity of proteins of various sizes against a variety of ligands. Since structural integrity is important, complete domains need to be displayed. The company is convinced that display and engineering of small stable proteins by this method will exhibit equivalent levels of diversity, and higher affinity and specificity, than with antibodies, and the company has a patent application on the technology.

3.6 MOLECULAR DIVERSITY OF PEPTIDES

Soluble Synthetic Peptide Combinatorial Libraries (Houghten Pharmaceuticals Inc., La Jolla, CA)

Since the introduction of methods for creating random peptide libraries in phage as a source of specific binding molecules (Devlin et al. 1990), researchers have searched for analogous soluble methods to create peptide libraries. The SPCL procedures are novel in that they allow peptide mixtures to be generated free in solution, thus allowing them to interact with their target receptor using conventional assay systems (Houghten et al. 1991). Further SPCL allows mixtures of free peptides to be generated, significantly increasing peptide diversity. The SPCLs are used in conjunction with an iterative selection process that uses positional-scanning synthetic peptide combinatorial libraries (PS-SPCL) to provide complete information about the peptide in a single assay. Modifications to the peptides to develop non-peptide libraries are also possible. SPCLs can be used in existing *in vitro* assay systems such as enzyme-linked immuno-sorbent assays (ELISAs), and allow unusual amino acids to be utilized (Pinilla et al. 1992; Benner 1994).

The Houghten procedure considers amino acid 6-mers with acetylated N-terminals and amidated C-terminals where the first two positions are specifically chosen from one of the 20 L-amino acids and the next four positions can be any mixture (combination) of monomers comprising L-, D-, or "unusual" amino acids (chemically modified amino acids). Then there are at least 64,000,000 possible 6-mers, and the SPCL process can screen and select through each of these to find a unique 6-mer with the desired specificity in 10 cycles, typically. (The process is illustrated in Fig. 3.7.)

An ELISA plate coated with a 6-mer of the form (OO-XXXX), where O is chosen from the 20 standard amino acids (except cysteine or tryptophan) and X is any randomly chosen amino acid, is exposed to an antibody that has been raised to that peptide in competition with a larger, 13-residue peptide. The one that causes the greatest inhibition of antibody binding is selected and 20 new peptides with each of the amino acids at position 3 now added to generate a 6-mer of the form (OOO-XXX). This iterative process is carried out for each of the remaining positions along the 6-mer until a 6-mer with an IC_{50} in the micromolar range is found. (IC_{50} here is the concentration of peptide need to inhibit 50% of antibody binding to the control peptide on the ELISA plate.) The two advantages of the methodology are that they allow free peptide to be assayed (hence any conventional bioassay may be used) and that the relative specificity of each position along the peptide may be determined.

The technology has been applied to developing lead antimicrobial peptides against *Staphylococcus aureus, Escherichia coli,* and *Pseudomonas aeruginosa;* as well as the yeast *Candida albicans.* A single 6-mer was found to be effective against these microorganisms with minimum inhibitory concentrations (MIC) of the order 3.2–6.5 μg/ml. It has also been applied to generating peptides as opiod receptor antagonists that can competitively inhibit enkephalins in crude rat brain homogenates. The opiod μ receptor was chosen, since it is generally associated with pain relief (e.g., thermal pain) as well as regulating nonanalgesic effects such as respiratory depression, antidiuresis, immune system suppression, and physical dependence. The opiod κ receptor, by contrast, is most potent in the mediation of analgesia in response to pain induced by chemical stimuli. Developing molecules specific for the μ receptor that do not affect the kappa receptor, and vice versa, is thus important. Starting with an SPCL comprised of over 52 million peptides, iterative enrichment resulted in the discovery of a peptide sequence capable of displacing the enkephalin peptide (the so-called DAMGO peptide) in competitive receptor binding studies. Of particular importance was the presence of the N-terminus acetyl group for binding. Using D-amino acid monomers in a similar study resulted in a peptide with an IC_{50} in the nanomolar range (Dooley et al. 1995). The expectation of the Houghten procedure is that it will reliably generate peptides of six residues or longer that will be able to competitively bind their targets in the nanomolar range (see Fig. 3.2). This should make it particularly effective at developing peptides against protease molecules, such as viral proteases needed to process viral polyproteins into mature viral molecules (e.g., HIV protease).

Chemical Synthesis of Peptides (Affymax, Inc.)

Affymax has a proprietary technology known as very large-scale immobilized polymer synthesis (VLSIPS). VLSIPS harnesses photolithographic techniques developed in the semiconductor industry, solid-phase chemistry, and photolabile protecting groups to achieve light-directed, spatially addressable, parallel chemical synthesis of diverse chemical compounds (Jacobs and Fodor 1994). The primary advantage of combinatorial combinations of peptides arrayed on a light-sensitive surface is that they are spatially indexed, allowing up to 64,000 molecules to be assayed and binders detected in a single pass. Affymax has searched for peptide inhibitors to the protease thermolysin by inserting 10 amino acids combinatorially at two positions in a pentapeptide substrate, thus generating 100 different compounds spatially arrayed on a light-sensitive silicon chip. After acetylating the peptides, the array was exposed to thermolysin in solution for 10 minutes, washed, and the liberated amines from the digested peptides were labeled with fluorescein isothiocyanate (this process is illustrated schematically in Fig. 3.8). The fluorescently tagged peptides attached to the chip were detected with a scanning fluorescence microscope and the data plotted as photon counts. A 3D graph of

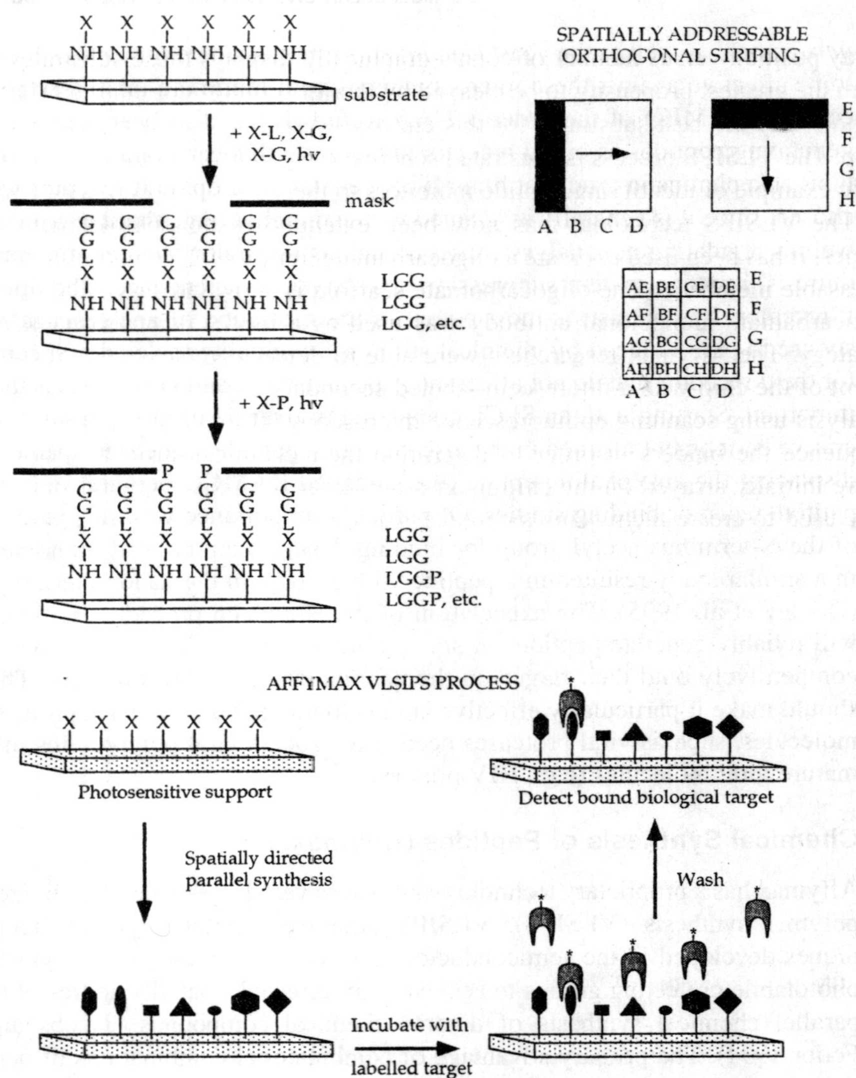

Figure 3.8. Spatial light-directed peptide synthesis (Affymax VLSIPS process). In the Affymax process nascent peptides (or small molecules) are synthesized on an immobilized light-sensitive support, where the order of the peptide sequence is determined by placement of masks over the support that direct the addition of the next amino acid (X–L, X–G, X–P, etc., in the figure) via the addition of a quantum of light ($h\nu$). Using orthogonal striping of the masks, the nascent peptide sequences can be identified to a unique spatial location on the grid (peptide LGGR at position AE, peptide LGGP at position BE, etc.). The peptides are then incubated with a labeled (either radioactive or fluorescent) target, washed, and the bound biological target used to identify the spatial location of the peptide with the strongest affinity. A simple lookup then identifies the sequence of this peptide.

array position versus number of counts graphically displays those substrates with the greatest propensity to be cleaved by thermolysin (Fodor et al. 1991). In this way the best substrates for this enzyme were determined in a single step. The VLSIPS process is illustrated schematically in Fig. 3.8 and the specific example of identifying peptide inhibitors to thermolysin in Fig. 3.9.

The VLSIPS technology has now been extended to nonpeptide backbones. It has been used to create a oligocarbamate library of 256 spatially addressable members. (The oligocarbomate scaffold is shown in Fig. 3.3). An anticarbamate monoclonal antibody was used as a model receptor in this strategy. The Affymax researchers were able to detect complexes by treatment of the chip with a fluorescein-labeled secondary antibody followed by analysis using scanning epifluorescence microscopy. To obviate the need to sequence the binders in order to determine their chemical structure, they were initially arrayed on the chip in an order defined by the synthetic strategy used to create them. This technique, known as *binary masking,* allows

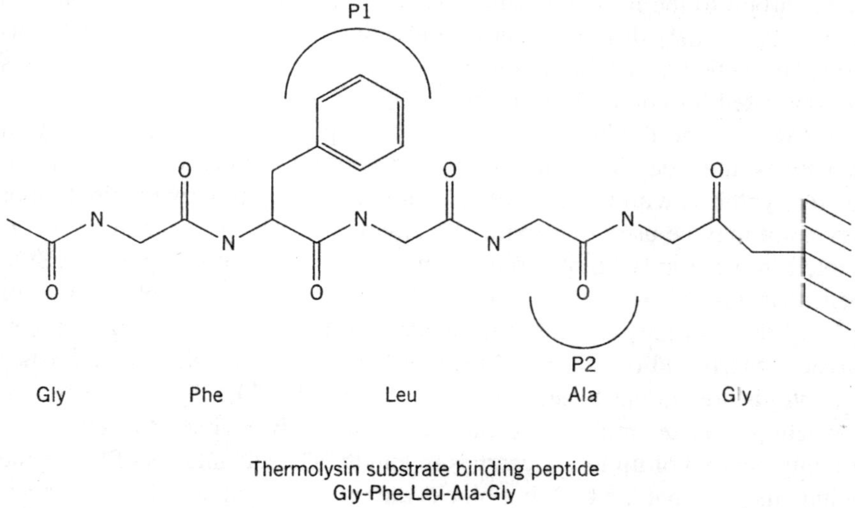

P1

P2

Gly Phe Leu Ala Gly

Thermolysin substrate binding peptide
Gly-Phe-Leu-Ala-Gly

Figure 3.9. Combinatorial exploration of protease specificity. The VLSIPS procedure described in the previous figure has been used to explore the peptide specificity of the protease thermolysin. Starting with the known substrate to the protease thermolysin (a suitable test case for other therapeutically relevant proteases), arrays of the peptide sequence Ac-Gly-R1-Leu-R2-Gly-NH were created as described in the previous figure (R1 and R2 are monomers chosen from an amino acid shape library including nonstandard and "unusual" amino acids. In the figure, R1 is Phe and R2 is Ala.). These spatially oriented, immobilized arrays were then digested with thermolysin and the liberated amines tagged with a fluorophore and identified. Location of their spatial position identifies the sequence of those peptides that are the best substrates for thermolysin.

researchers to backtrack to the chemical structure of the binders simply from their position on the array. Binary masking is illustrated schematically in Fig. 3.8.

Drug Discovery Using oligo N-Substituted Glycines ("Peptoids") (Chiron Corp.)

The development of solid-phase synthesis of peptides and oligonucleotides combined with mixing strategies has now allowed the generation of libraries of tens of thousands of compounds per experiment. However, the metabolic instability of the peptide and oligonucleotide monomer units by proteolysis or nuclease activity and the poor absorption characteristics of amino acids and nucleic acids limits their role as drug candidates. Chiron Corp. has developed a procedure using chemically modified peptides that can still be easily synthesized stepwise using solid phase chemistry. Chiron's "NSG peptoids" are N-substituted glycines (the side chains are moved from the alpha-carbon to the nitrogen) with a series of different side chains that can be chosen to satisfy different pharmacokinetic properties (Farmer and Ariens 1983; Horwell et al. 1987; Simon et al. 1992) (peptide and peptoid scaffolds are illustrated for comparison in Fig. 3.10).

Because of the modular approach to solid-phase synthesis, a variety of backbones and side chains are possible in peptoid synthesis. Several methods for synthesis with these monomers are possible. Reductive amination or alkylation may be used to provide the monomers, which can then be incorporated into the growing peptoid chain using the activating agents PyBOP (benzotriazol-1-yloxtris (pyrrolidino) phosphonium hexaflurophosphate) or PyBroP (bromotris (pyrroloidino)-phosphonium hexaflurophosphate) under classical Merrifield conditions (Merrifield 1963). Yields and characteristics of peptoids are similar to those of peptides. Alternatively, submonomers of alternating acetate/amine units can be sequentially reacted to generate a growing peptoid of up to 25 residues in length. The advantages of this route include the fact that the building blocks are primary amines, which are commercially available in large quantities and are functionally diverse. The solid-phase synthesis of peptoids is illustrated in Fig. 3.10.

Optimization of combinatorial libraries from the 10^{12} tetramers possible from 1000 readily available primary amines means that massive libraries may be quickly synthesized. It would take millions of years to screen such a library *a posteriori;* hence methods are needed to select specifically which diverse libraries should be created to maximize the opportunity of finding a lead and which related molecules should be generated to optimize that lead. Optimization can be achieved by ensuring that the first round of oligomers

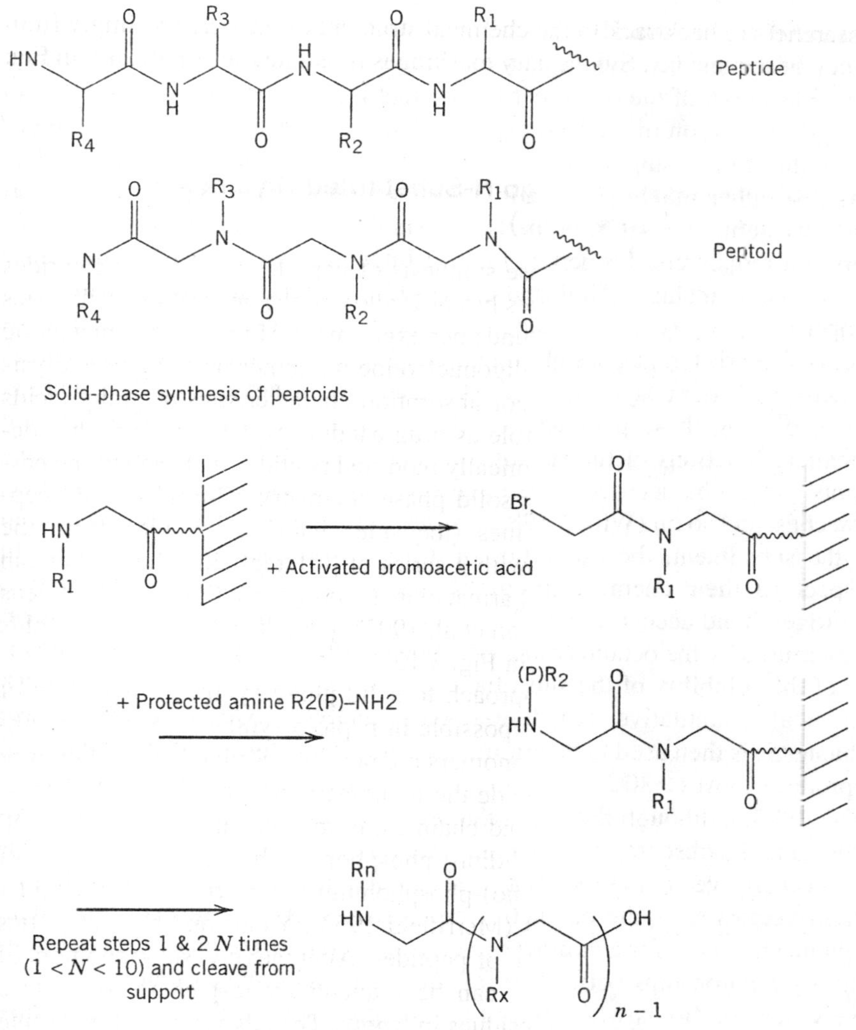

Peptide

Peptoid

Solid-phase synthesis of peptoids

+ Activated bromoacetic acid

+ Protected amine R2(P)–NH2

Repeat steps 1 & 2 N times ($1 < N < 10$) and cleave from support

Figure 3.10. Solid-phase synthesis of peptoids. Peptoids resemble peptides except that the side chains (R-groups) are moved from the alpha-carbon to the nitrogen. Because of the commercial availability of a large number of primary amines, solid-phase synthesis of peptoids with a wide variety of shapes, branching, and functional groups is possible. In the Chiron process, the monomer pool of amines are first analyzed using 16 different properties to determine their molecular diversity; those that are sufficiently diverse to represent the library from which they originate are then selected for synthesis of up to 10-mer length peptoids. Nanomolar range inhibitors to some therapeutically relevant targets have been discovered starting with initial pools of 5000 peptoids.

generated are maximally dissimilar to each other, minimizing the redundancy of the test set. Synthesis then focuses on oligomer libraries generated from the subset of the monomer library that is most similar to the side chain at a given position in the lead peptoid. These optimization strategies require the ability to measure similarity between potential monomers and to select sets that either maximize or minimize the resemblance. Similarity is measured by defining a set of property vectors that define the similarity distance between monomers. Properties include lipophilicity (Leo 1993), molecular shape and branching (Hall and Kier 1991), functional groups, and atom properties. Five types of shape descriptors were used as topological indices of the different possible functional groups that may be added. Three-dimensional atom layering was also used, and a novel graphical method for displaying these parameters ("flower plots") developed. The different chemical functions of the substituents are represented by database "fingerprints" that characterize substituents by the presence of rings, chains, branches, and so on (Willett 1987). Atom "layering" tabulates, for each atom in the substituent, the properties of all atoms in successive "shells" with respect to their chemical properties (acid, base, hydrogen-bond donor, hydrogen-bond acceptor, etc.) The overall hydrophobicity of the substituent is measured by the octanol/water partition coefficient log P (the log of the ratio of the solubility of the substituent in octanol and its solubility in water). Classical quantitative structure-activity analysis (QSAR analysis; see Chapter 5) is then used to determine the number of principal components that reproduce most ($>80\%$) of the observed variance in the properties (Martin et al. 1995). Although the exact chemical significance of any of these properties may be obscure, it is generally observed that compounds sharing similar property vectors are similar in shape and affinity, while those with different vectors are quite distinct. (These metrics have been used in physical organic chemistry for a number of years to quantitate the observed structure-activity relationships between different functional groups of atoms on relatively inert scaffolds when placed in biological systems; this subject is discussed in more detail in the QSAR section of the next chapter.)

The goal of having a series of metrics that accurately describes the physical-chemical properties of molecular fragments is to enable the best or optimal choice of substituents for chemical synthesis. Once a series of metrics have been determined that accurately parameterizes molecular similarity, it can be used in computer-based chemical database searches to find chemical compounds that contain identical or similar functional groups that can be used to optimize monomer diversity. Statistical techniques such as partial least squares analysis and multidimensional scaling techniques can be used to scan "fingerprints" from substituent group databases and then predict their properties by QSAR. Chiron uses a total of 16 properties to describe each monomer. The monomer "pool" is then ranked by distance metric from the reference target monomer. To find reasonably dissimilar sets (in part to satisfy robotics limitations in the number of molecules that can be synthesized), a procedure called D-optimal design is used. This ensures that the molecules synthesized contain monomers that have known pharmacophores for the receptor yet are diverse from both the initial set and from each other. (The concept of pharmacophore development is discussed in the next chapter.) The D-optimal algorithm chooses points laid out in property space that are well spread out and (approximately) orthogonal, representing substituents that are molecularly diverse. After user selection of a minimum number of monomers from the pool, the optimization algorithm selects other points for inclusion in the set that best encompass the diversity of the substituents.

One advantage with peptoids is that they are expected to improve membrane permeability (absorption) and *in vivo* lifetime (resistance to proteolytic degradation), since the amide bond most susceptible to proteolysis has been N-substituted. Peptoids are expected to be particularly useful in cases where peptide therapeutics are fallible, typically where stability and oral availability are a concern. As a validation of this strategy, Chiron has discovered a 7-nM nonpeptide inhibitor to the μ-opioid peptide receptor and an 8-nM inhibitor to the alpha-1 adrenergic receptor from an initial pool of 5000 peptoids (Martin et al. 1995).

3.7. MOLECULAR DIVERSITY OF OLIGONUCLEOTIDES

Randomized Oligonucleotide Libraries (Isis Pharmaceuticals, Inc., Gilead Sciences, Inc.)

Built upon the specificity of Watson-Crick base-pairing, antisense oligonucleotides are short stretches of DNA or RNA that recognize and hybridize to a specific complementary gene sequence or messenger RNA molecule and inhibit their action by physically blocking the template sequence. For oligonucleotides to be used in the therapy of human disease, however, they must be easily synthesizable in bulk, be stable *in vivo,* be able to enter the target cell, be retained by the target cell, be able to interact with their (cellular) targets, and not be able to interact with other macromolecules. Stability of antisense oligonucleotides can be engineered by chemical modification to the sugar-phosphate backbone, such as by addition of a thiol group to the phosphate. Evaluation of phosphorothioate (PS) oligonucleotides suggests both sequence-dependent and sequence-independent binding to target genes, and thus that the ultimate therapeutic value of these molecules depends on determining their therapeutic index (therapeutic potential) in humans. Nonetheless, antisense therapeutics have tremendous potential, particularly since their design is conceptually trivial once a known disease gene has been isolated and at least partially sequenced.

A development in combinatorial libraries of antisense oligonucleotides has been the discovery that certain oligonucleotides will form triplexes when complexed with double-stranded DNA or RNA. Oligonucleotides that achieve this can inhibit transcription from these sites by preventing the local unwinding of the double helix that is required for transcription factor proteins to bind. The affinity of antisense oligonucleotides for their nucleic acid targets depends on hydrogen bonding and base stacking, and consequently increases with length of the oligo-receptor complex. Affinity also varies with the sequence of the duplex. The specificity of these molecules and much of the initial excitement over their use as therapeutic drugs emerges from the selectivity of Watson-Crick (and other, e.g., Hoogstein) base-pairing. Simply by synthesizing an oligo with sequence complementary to that of the target, selectivity is ensured. Statistical analyses suggest that oligonucleotides of length 12–30 bases are least likely to crosshybridize to nontarget stretches of DNA in the host cell. Similarly oligonucleotides of slightly shorter length, 11–15 bases, are required to select out specific RNA species. The decrease in affinity (and selectivity) due to mismatch base-pair recognition has been studied and found to focus on the specific mismatch, the position, and the region of complementarity surrounding the mismatch. Initial *in vivo* studies of antisense oligos have shown activity against herpes simplex virus I infection, tick-borne encephalitis virus, and HIV-I. An antisense oligo to the *c-Myb* gene has also been shown to be effective in a mouse SCID model of human leukemia (U. of Pennsylvania and Lynx Pharmaceuticals, Inc.).

ISIS Pharmaceuticals has developed a technique called SURF (Synthetic Unrandomization of Randomized Fragments) to screen phosphorothioate oligonucleotide libraries containing 4^8 (65,536) octamers divided into 16 sets of 4096 compounds each, screened against a cell-based assay of HIV-mediated cell fusion. Subsequent rounds of screening and selection showed that four consecutive guanosines were required for maximum antiviral activity. No strong selection preference was observed for nucleotides flanking the guanosine core, however. The compound T2G4T2 (TTGGGGTT) was chosen for further study, and organic mimetics of this molecule are under development as possible clinical compounds. The oligonucleotide T2G4T2 exhibits good hybridization properties to its target, is nuclease resistant, and shows good pharmacokinetics in animal and human models as well as acceptable toxicity levels in early human trials. This molecule seems to work by hybridizing to the V3 loop of the HIV envelope protein gp120, inhibiting the virion from cell fusion, and preventing cell-to-cell transmission of HIV and concomitant syncytia (plaques of infected cells) formation. In addition T2G4T2 is additive or synergistic with AZT, suggesting its use as a virucidal agent during initial infection as well as during full-blown AIDS.

ISIS has extended their combinatorial oligonucleotide synthesis to chemically modified nucleotide analogs that provide an extended array of functionalities such as Hoogstein (an alternative method of base-pairing) as well as Watson-Crick base-pairing, ion-pair formation, hydrophobic and aromatic interactions, and dipole-dipole interactions. Synthetic properties desired of these novel backbone structures include compatability with automated synthesis techniques, efficient preparation, controlled chirality, and patentability. Examples of backbones include trans-4-Hydroxy-L-proline (HPP), (R)-(+)-Glycidol (EGP), and peptide nucleic acids (PNAs). Creation of 12-residue oligomers with complete randomization at 3 or 4 positions detected micromolar inhibitors to phospholipase A2. ISIS is confident their non-oligonucleotide extensions to SURF will enable them to discover nanomolar inhibitors to this and other enzymes by providing controlled oligomerization with unlimited structural diversity.

Gilead Sciences, Inc. uses its "aptamer" technology to create vast libraries of oligonucleotides (up to 10^{13} in a single run) up to 50 bases in length, and exposed them to thrombin and other proteins a target. Tight-binding molecules were PCR amplified and used as templates for a further round of selected mutagenesis, screening, and amplification. In each pass only about 0.01% (10^9) molecules are found to bind; however, five generations of this form of directed molecular evolution resulted in a potent thrombin inhibitor currently being tested in animals as an anticoagulant. Although thrombin is not a DNA-binding protein, it is clear that this process is able to generate sufficient shape diversity within an oligonucleotide pool to select and enrich an aptamer with the needed conformational characteristics (three-dimensional shape complementarity) to bind the thrombin receptor site with high specificity and affinity.

Molecular Diversity of Ribonucleic Acid (RNA) Ligands (U. of Colorado, Nexagen, Inc.)

A procedure has been developed to screen a very large combinatorial library of RNA ligands against a variety of targets (Tuerk and Gold 1990; Gold et al. 1993). The procedure, termed SELEX (Systematic Evolution of Ligands by EXponential enrichment) has generated high affinity RNA ligands for several targets, including nucleic acid binding proteins (Tuerk and Gold 1990; Tuerk et al. 1993); other binding proteins (Bock et al. 1992, 1993); and small molecules (Sassanfar and Szostack 1993; Jenison et al. 1994). Further the SELEX procedure has been shown to be able to identify ligands that can distinguish between different, though related, targets such as the reverse transcriptases of different retroviruses (Chen and Gold 1994).

The SELEX procedure begins with a starting repertoire containing up to 10^{14} unique RNA species to which target proteins are bound. Bound RNA is partitioned by nitrocellulose filtration, recovered, and used as a template for cDNA synthesis. The cDNA product is then amplified by PCR and transcribed with T7 RNA polymerase to generate the RNA pool for the next round of selection. The starting repertoire typically consists of one or more randomized regions separated by fixed nucleic acid sequences flanked by primer binding sequences. Sequential rounds of selection, amplification, and mutagenesis leads from starting ligands with micromolar range affinities to a 10^3 increase after 10–12 rounds of selection, resulting typically in 10–50 ligands with low nanomolar affinities. Secondary structure predictions of the best binders suggests that they form well defined structures such as stem/loops and pseudoknots (Zuker 1989). Indeed, beginning with a starting repertoire containing consensus sequences for pseudoknot or stem/loop formation assists the SELEX procedure in obtaining RNA ligand structures with the requisite affinities for the target.

MD has also been applied to RNA molecules that can cleave themselves or other RNA molecules. These molecules, initially discovered over 10 years ago in the single-celled ciliate *Tetrahymena,* are now known as ribozymes and are speculated to be the earliest molecules capable of self-replicating at the onset of life on earth. Their structures typically contain stretches of complementary RNA bases that can base-pair with a target RNA ligand (including the RNA molecule itself), giving rise to an active site of defined structure that can cleave the bound RNA molecule. Ribozymes have significant promise as therapeutics for diseases such as AIDS, cancer, and chronic hepatitis because they can be targeted against the messenger RNA of the invading organisms responsible for these diseases and thus prevent their replication within host cells. It has been established that reactions mediated by catalytic RNA molecules are fundamental to the biochemical function of a number of different organisms, and as such they serve as attractive candidates for interventional therapy. Bound RNA molecules that may serve as therapeutic targets might include viral RNAs made by virus-infected host cells or the mRNA product of a mutated oncogene. For instance, a "hairpin" ribozyme against an HIV consensus RNA sequence has been shown to reduce the amount of virus in HIV-infected cells 10,000-fold in cell cultures. The ribozyme specificity to the target RNA sequence should ensure minimal side effects, but these have still to be proved *in vivo.* (Ribozyme therapy will require effective delivery of the ribozyme to the appropriate part of the cell.) *In vitro* directed evolution has also been used to create ribozymes with entirely new catalytic functions such as RNA ligase activity (Ekland and Bartel

1995). The results of these experiments suggest that ribozymes might be designed for a wide variety of therapeutic applications. Further, through the use of DME, ribozymes can be created with novel catalytic specificities that can mediate aberrant biochemical pathways in diseased cells.

3.8. MOLECULAR DIVERSITY OF SMALL ORGANIC MOLECULES

Screening and Identification of Organic Compounds through Peptide Tags (Selectide, Inc., Tucson, AZ)

Selectide has developed a process that can tag hundreds of thousands of organic molecules by immobilizing them on beads and tagging them with labeled peptides. (This process is illustrated in Fig. 3.4.) Initial leads discovered by the Selectide process can be improved from micromolar to nanomolar affinity by synthesizing and screening secondary combinatorial libraries generated after identification of a tagged lead compound. The Selectide process allows the detection of binding of macromolecules (e.g., antibodies or other proteins) by flow cytometry of the molecules against compounds immobilized on beads or cells, or by a sandwich assay that measures fluorescence, radioactivity, or colorimetry between the binding surface of an immobilized lead and a soluble target. Because the Selectide process uses nonsequenceable organic compounds as leads in high-throughput screening, an easy to sequence (nucleic acid or amino acid) leader must be used to "tag" the molecules for subsequent identification and isolation. The process utilizes manual or automated solid-phase synthesis of libraries of test compounds. In the case where the support is beads, the assay of the lead compound against the target is achieved in such a manner that the parent bead for each positive lead can be recovered. The "split" synthesis method (Furka et al. 1988; Lam et al. 1991) is used for library synthesis. Positive compounds composed of nonsequenceable building blocks are attached to a coding sequence in such a way that the latter does not interfere with binding to the target. Either the screening compound is attached separately to the support which may be (either columns or beads) from the coding moiety (conditions under which the screening compounds are released are different than for the coding compounds) or the coding sequence is located in the interior of the solid support and is inaccessible to the target molecule. The latter process has shown to be successful whether the targets are enzymes or the soluble extracellular domains of transmembrane receptors. The process allows secondary libraries to be generated in days once initial compounds have been identified through the use of peptide tags; these can be used to identify a series with a distinct SAR (structure activity relationship), which can in turn be used to condition the library and improve activity. Occasionally nanomolar leads are optimized by the Selectide process, obviating the need to synthesize single compounds. However, there is no *a priori* reason for the compounds found to be optimally active to a particular target, and indeed the strength of the Selectide process is that it optimzes compounds to a range of targets rather than maximally optimizing a single compound for any particular target. Selectide is testing nanomolar-range affinity compounds against the IIb/IIIa receptor (found on the surface of platelet cells) and the blood protein thrombin in animal models. Both of these compounds were generated from initial micromolar leads.

Organic Diversity by "Diversomer" Synthesis (Warner-Lambert, Ann Arbor, MI)

A combination of solid-phase chemistry, organic synthesis, and multiple, simultaneous synthesis in specially constructed apparatus have been developed to allow the synthesis of organic compounds over two or three steps incorporating chemically diverse building blocks—called *diversomers*. The approach has been successfully used to synthesize dipeptides, hydantoins, and benzodiazepines. In the diversomer method, target compounds are simultaneously, but separately, synthesized on a solid support in an array format to generate an ensemble of structurally related compounds. Building blocks are sequentially coupled to the growing chain on a resin until the penultimate product at each location on the array is complete. Cleavage from the resin yields a final product that can readily be separated from the spent resin. Typically 40 to 50 compounds can be synthesized and screened by the process. The compounds can then be screened in a competitive binding assay to ascertain their biological activity. Criteria for useful diversomer synthesis include the ability to simultaneously, but separately, synthesize related compounds in soluble format, sufficient quantity (>1 mg), and purity; automation for speed, accuracy, and precision and utilization of a synthetic strategy that is compatible with general synthesis procedures. The intermediate products of the diversomer process are readily separable from the byproducts and excess reagents. The Warner-Lambert strategy satisfies these criteria but lacks the ability to randomly modify the growing polymers and amplify those that show potential and positive lead compounds.

The goal of the diversomer method is to explore a larger region of SAR (structure activity relationship) space than has been obtainable by traditional methods. Since a large SAR set already exists for the drugs Dilantin (a hydantoin) and Valium (a benzodiazepine), combinatorial synthesis of novel structural analogs by the diversomer procedure is a good test of the value of the procedure in identifying and optimizing design strategies for these drug molecules. Anticipating the success of this approach, Warner-Lambert has recently spun off the diversomer technology into a separate business entity.

3.9. SUMMARY

Drug discovery and medicinal chemistry is ultimately about finding small stable organic molecules that bind to a desired target in the body with extraordinary affinity. Molecular diversity and combinatorial chemistry promise to create millions of compounds simultaneously that can explore the complex surface of the property space parameterized by molecular shape, charge, force, or any other characteristic that may be important to binding to a target. The odds of finding a compound with the required affinity (and without side effects such as toxicity or instability) are dramatically increased when millions can be created that together encompass as large a region of property space as possible.

Although there are a multitude of different methodologies for generating molecular diversity, all require the ability to assemble every possible combination of a given set of molecular building blocks, while simultaneously recoding which have been used (and in what order) and then assaying the resulting compounds simultaneously and selecting from the record those that seem promising. Two main subdivisions among companies following this strategy can be seen: those using "mutatable" molecules (peptides and oligonucleotides) and the process of directed molecular evolution to rapidly optimize promising molecules that can then act as templates for the next round of optimization, and those using small organic molecules as building blocks, which, though they cannot be mutated, can exhibit a range of properties resulting in diversity that is considerably greater than available by using only biological polymers. (A third strategy is to create and refine large numbers of peptides or oligonucleotides until a nanomolar-range molecule is found, and then to convert that "lead" into a small organic compound using the drug design methodology described in the next chapter.) For organic molecules to be identified, an oligonucleotide or peptide linker can be used that serves as a molecular "bar code" to the compound. (Leads based on peptides or oligonucleotides use the sequence inherent in the polymer as its bar code.) One way companies using organic molecules can simulate directed evolution is by adding new chemistries (new reactions) that are randomly selected to determine the coupling of one monomer to the next. This method can be coupled with a screening procedure, allowing one to iterate toward

METHODS FOR QUANTITATING MOLECULAR DIVERSITY IN COMBINATORIAL LIBRARIES

The generation and quantitation of the molecular diversity in a combinatorial library can be measured by a number of methods.

1. *Diversity metrics.* These measure the diversity in a compound library and may suggest areas where the library is either over- or underrepresented. Metrics include the classical measures of molecule substituent physical-chemical properties, such as the Hansch-Leo substituent constants log *P* (the partition coefficient of a molecule in octanol and water), sigma, and pi; (measures of substituent bulk and hydrophobicity) the Swain-Lupton F and R values of inductive and polarizable electronic distribution, respectively; or the Verloop B-values of steric bulk. More sophisticated metrics of molecular features might include measures of 3D QSAR such as CoMFA fields (described in the next chapter), molecular similarity, or molecular shape and connectivity (e.g., atom layering). Two-dimensional and three-dimensional "fingerprints" of molecules or molecular fragments provide a relatively simple way of quantifying diversity using validated metrics. (See the section on the Chiron process for quantifying diversity in peptoid libraries.) Since many of these diversity metrics overlap, it is important to determine which ones are orthogonal and contribute the most to the quantitation of structural diversity. The statistical technique of principal components analysis (PCA, described in Chapter 5) can be used to "boil down" metrics to a handful of measures that quantitate the diversity in a given substituent list. The validation of diversity metrics against measured activities is expected to be essential to the usefulness of this approach.
2. *Clustering methods.* These use standard measurements of molecular shape or connectivity to cluster together related molecular structures from which a diverse subset may be chosen. A number of clustering algorithms exist, such as hierarchical, Jarvis-Patrick, and reciprocal nearest-neighbor, each with its own strengths and weaknesses. Clustering algorithms attempt to determine what properties of molecular substituents are truly orthogonal, that is, measure significantly different aspects of the structure of the molecules. The "similarity principle" proposes that structurally similar functional groups will exhibit similar physicochemical and biological properties when tagged onto molecular scaffolds and that the extraction of

a cluster (either by creating a combinatorial library or by searching an existing library) will lead to a prioritized list of targets to be synthesized. Conceptually clustering can be divided into three phases: characterization of the set in terms of descriptors, calculation of pairwise similarity between members of the set, and grouping of members based on the distance calculations. Hierarchical clustering methods allow the construction of dendrograms ("trees") consisting of nested sets, which allow one to visually select out representative members for prioritized synthesis. Alternatively, nonhierarchical clustering (e.g., Jarvis-Patrick) creates nearest-neighbor lists with either larger looser clusters or smaller tighter clusters (dependent on a user-settable parameter). In either case the purpose of clustering algorithms is to determine which representative compounds or functional groups are chosen for synthesis or pooling from the large numbers of potential candidates.

In reality, medicinal chemists often have a lead compound or lead hypothesis so that combinatorial synthesis will proceed from this model. Quantitation of diversity from a substituent list and the clustering of functional groups is essential in creating the strategy for synthesis of the compounds, since the experimenter will need to determine precisely which compounds or mixtures are placed into which well of each microtiter plate. In the strategy of pooled synthesis, it is preferred that compounds that are maximally dissimilar not be diluted into single wells, since then the likelihood of detecting a tight or even moderate binder in any given well is substantially lessened. Since the thrust of combinatorial chemistry is to array a variety of scaffolds or templates with the appropriately chosen substituents, selecting the appropriate mixtures is essential to the drug discovery process.

pharmacologically more active compounds. In theory, combinatorial chemistry should provide a formidable data set for SAR studies, yet in practice so much data are generated that, without a significant automated chemical information management system in place, a company may well be overwhelmed by the data they generate. Issues in any combinatorial chemistry program that companies need to address include: What library design strategies provide the best libraries? How do metrics for measuring library diversity scale with the size of the libraries? Do metrics that measure diversity in molecular structure also measure diversity in biological function? Can library diversity be focused on pharmacologically valid molecules or targets?

Ultimately the greatest benefit that molecular diversity can bring is in the ability to produce sufficient data and so-called lead molecules for true molecular design by the techniques described in the following chapter.

REFERENCES

Bass, S., Greene, R., and Wells, J. A. 1990. Hormone phage: An enrichment method for variant proteins with altered binding properties. *Proteins* 8(4):309–314.

Bock, L. C., Griffin, L. C., Lantham, J. A., Vermaas, E. H., and Toole, J. J., 1992. Selection of single-stranded DNA molecules that bind and inhibit human thrombin. *Nature* 355(6360):564–566.

Bunin, B. A., Plunkett, M. J., and Ellman, J. A. 1994. The combinatorial synthesis and chemical and biological evaluation of a 1,4-benzodiazepine library. *Proceedings of the National Academy of Sciences* 91(11):4708–4712.

Burbaum, J. J., Ohlmeyer, M. H., Reader, J. C., Henderson, I., Dillard, L. W., Li, G., Randle, T. L., Sigal, N. H., Chelsky, D., and Baldwin, J. J. 1995. A paradigm for drug discovery employing encoded combinatorial libraries. *Proceedings of the National Academy of Sciences* 92(13):6027–6031.

Chen, H., and Gold, L. 1994. Selection of high-affinity RNA ligands to reverse transcriptase: Inhibition of cDNA synthesis and RNase H activity. *Biochemistry* 33(29):8746–8756.

de la Cruz, V. F., Lal, A. A., and McCutchan, T. F. 1988. Immunogenicity and epitope mapping of foreign sequences via genetically engineered filamentous phage. *Journal of Biological Chemistry* 263(9):4318–4322.

Devlin, J. J., Panganiban, L. C., and Delvin, P. E. 1990. Random peptide libraries: A source of specific protein binding molecules. *Science* 249(4967):404–406.

DeWitt, S. H., Kiely, J. S., Stankovic, C. J., Schroeder, M. C., Cody, D. M., and Pavia, M. R. 1993. "Diversomers": An approach to nonpeptide, nonoligomeric chemical diversity. *Proceedings of the National Academy of Sciences* 90(15): 6909–6913.

Dooley, C. T., Chung, N. N., Schiller, P. W., and Houghten, R. A. 1993. Acetalins: Opiod receptor antagonists determined through the use of synthetic peptide combinatorial libraries. *Proceedings of the National Academy of Sciences* 90(22):10811–10815.

Dooley, C. T., Kaplan, R. A., Chung, N. N., Schiller, P. W., Bidlack, J. M., and Houghten, R. A. 1995. Six highly active mu-selective opioid peptides identified from two synthetic combinatorial libraries. *Peptide Research* 8(3):124–137.

Ekland, E. H., and Bartel, D. P., 1995. The secondary structure and sequence optimization of an RNA ligase ribozyme. *Nucleic Acids Research* 23(16):3231–3238.

Erb, E., Janda, K. D., and Brenner, S. 1994. Recursive deconvolution of combinatorial chemical libraries. *Proceedings of the National Academy of Sciences* 91(24):11422–11426.

Farmer, P. S., and Ariens, E. J. 1993. *Trends Pharmacol. Sci.* 3:363–365.

Fodor, S. P., Read, J. L., Pirrung, M. C., Stryer, L., Lu, A. T., and Solas, D. 1991. Light-directed, spatially addressable parallel chemical synthesis. *Science* 251(4995):767–773.

Furka, J. A., et al. 1988. *14th Int. Cong. Biochem* 5:47.

Geysen, H. M., Meloen, R. H., and Barteling, S. J. 1984. Use of peptide synthesis to probe viral antigens for epitopes to a resolution of a single amino acid. *Proceedings of the National Academy of Sciences* 81(13):3998–4002.

Gold, L., et al. 1993. *The RNA World. RNA: The Shape of Things to Come.* Cold Spring Harbor; NY: Cold Spring Harbor Laboratories Press.

Hall, L. H., and Kier, L. B. 1991. In *Reviews in Computational Chemistry,* vol. 2, Lipkowitz and Boyd, (eds). VCH Publishers, pp. 367–421.

Horwell, D. C., Beeby, A., Clark, C. R., and Hughes, J. 1987. Synthesis and binding affinities of analogues of cholecystokinin-(30-33) as probes for central nervous system cholecystokinin receptors. *Journal of Medicinal Chemistry* 30(4): 729--732.

Houghten, R. A., Pinilla, C., Blondelle, S. E., Appel, J. R., Dooley, C. T., and Cuervo, J. H. 1991. Generation and use of synthetic peptide combinatorial libraries for basic research and drug discovery. *Nature* 354(6348):84–86.

Jacobs, J. W., and Fodor, S. P. 1994. Combinatorial chemistry-applications of light-directed chemical synthesis. *Trends in Biotechnology* 12(1):19–26.

Jenison, R. D., Gill, S. C., Pardi, A., and Polisky, B. 1994. High-resolution molecular discrimination by RNA. *Science* 263(5152):1425–1429.

Kang, A. S., Barbas, C. F., Janda, K. D., Benkovic, S. J., Lerner, R. A. 1991. Linkage of recognition and replication functions by assembling combinatorial antibody Fab libraries along phage surfaces. *Proceedings of the National Academy of Sciences* 88(10):4363–4366.

Lam, K. S., Salmon, S. E., Hersh, E. M., Hruby, V. J., Kazmierski, W. M., and Knapp, R. J., 1991. A new type of synthetic peptide library for identifying ligand-binding activity. *Nature* 354(6348):82–84.

Leo, A. 1993. *Chemical Review* 93:1281.

Lowman, H. B., Bass, S. H., Simpson, N., and Wells, J. A. 1996. Selecting high-affinity binding proteins by monovalent phage display. *Biochemistry* 30(45): 10832–10838.

Markland, W., Roberts, B. L., Saxena, M. J., Guterman, S. K., and Ladner, R. C. 1991. Design, construction and function of a multicopy display vector using fusions to the major coat protein of bacteriophage M13. *Gene* 109(1):13–19.

Martin, E. J., Blaney, J. M., Siani, M. A., Spellmeyer, D. C., Wong, A. K., and Moos, W. H. 1995. Measuring diversity: Experimental design of combinatorial libraries for drug discovery. *Journal of Medicinal Chemistry* 38(9):1431–1436.

McCafferty, J., Griffiths, A. D., Winter, G., and Chiswell, D. J. 1990. Phage antibodies: Filamentous phage displaying antibody variable domains. *Nature* 348(6301):552–554.

Merrifield, R. B. 1963. *Journal of the American Chemical Society* 85:2149–2154.

Parmley, S. F., and Smith, G. P. 1988. Antibody-selectable filamentous fd phage vectors: affinity purification of target genes. *Gene* 73(2):305–318.

Pinilla, C., Appel, J. R., Blanc, P., and Houghten, R. A. 1992. Rapid identification of high affinity peptide ligands using positional scanning synthetic peptide combinatorial libraries. *Biotechniques* 13(6):901–905.

Roberts, B. L., Markland, W., Ley, A. C., Kent, R. B., White, D. W., Guterman, S. K., and Ladner, R. C. 1992. Directed evolution of a protein: Selection of potent neutrophil elastase inhibitors displayed on M13 fusion phage. *Proceedings of the National Academy of Sciences* 89(6):2429–2433.

Sassanfar, M., and Szostak, J. W. 1993. An RNA motif that binds ATP. *Nature* 364 (6437):550–553.

Schultz, P. G., and Lerner, R. A. 1995. From molecular diversity to catalysis: Lessons from the immune system. *Science* 269(5232):1835–1842.

Scott, J. K., and Smith, G. P. 1990. Searching for peptide ligands with an epitope library. *Science* 249(4967):386–390.

Sepetov, N. F., Krchnak, V., Stankova, M., Wade, S., Lam, K. S., and Lebl, M. 1995. Library of libraries: Approach to synthetic combinatorial library design and screening of "pharmacophore" motifs. *Proceedings of the National Academy of Sciences* 92(12):5426–5430.

Simon, R. J., Kania, R. S., Zuckermann, R. N., Huebner, V. D., Jewell, D. A., Banville, S., Ng, S., Wang, L., Rosenberg, S., Marlowe, C. K., et al. 1992. Peptoids: A modular approach to drug discovery. *Proceedings of the National Academy of Sciences* 89(20):9367–9371.

Simon, R. J., Kania, R. S., Zuckermann, R. N., Huebner, V. D., Jewell, D. A., Banville, S., Ng, S., Wang, L., and Rosenberg, S. 1994. In *Techniques in Protein Chemistry V.* San Diego, CA: Academic Press.

Smith, G. P. 1985. Filamentous fusion phage: Novel expression vectors that display cloned antigens on the virion surface. *Science* 228(4705):1315–1317.

Stone, D. 1993. *Biotechnology* 11:1508.

Tuerk, C., and Gold, L. 1990. Systematic evolution of ligands by exponential enrichment: RNA ligands to bacteriophage T4 DNA polymerase *Science* 249(4968): 505–510.

Tuerk, C., MacDougal, S., and Gold, L. 1992. RNA pseudoknots that inhibit human immunodeficiency virus type 1 reverse transcriptase. *Proceedings of the National Academy of Sciences* 89(15):6988–6992.

Willett, P. 1987. In *Similarity and Clustering in Chemical Information Systems.* Wiley, New York: p. 54.

Zuker, M. 1989. On finding all suboptimal foldings of an RNA molecule. *Science* 244(4900):48–52.

4

PROTEIN ENGINEERING AND COMPUTER-ASSISTED DRUG DESIGN

4.1. AN OVERVIEW OF THE ROLE OF COMPUTATIONAL CHEMISTRY IN THERAPEUTIC DRUG DESIGN

Computational drug design arose with the realization that computers were beginning to get powerful enough to allow the simulation of the physicochemical properties of drug molecules and their receptors. The drug industry's need to develop products systematically, together with the inefficiency of classical drug discovery, has led to the progressive development of the technique of "rational" drug design—the use of computers to literally design drugs atom by atom. Since the first developments over 20 years ago, computational drug design has expanded to include new compound discovery (by computer searching of chemical databases), compound optimization (the systematic modification of functional groups to maximize potency and minimize or eliminate side effects such as toxicity), and even *de novo* drug design (the ability to generate entirely new molecules that might fit a receptor site and act as specific antagonists or inhibitors). Formally computational chemistry is the quantitative modeling of chemical behavior on a computer by the formalisms of numerical methods. This includes database systems that can search for structures, activities, or properties; tools for analyzing experimental data such as those from spectroscopic or diffraction studies; modeling and visualization systems for examining and predicting chemical properties and structures; and computer-assisted synthesis and planning systems that propose synthesis or reaction pathway schemes for compound development. Molecular modeling itself encompasses the generation and representation of the three-dimensional structure of molecules and their physicochemical properties. This includes structure building and conversion from one or two-dimensions (2D) into three-dimensions (3D), simulation of chemical properties and behavior, analysis of structures and their associated properties, and quantitative methods to compare chemical structures for similarities or differences that may be related to their physical properties. While not every pharmaceutical or biotechnology company has embraced all aspects of computational drug design, virtually all utilize computers in some part of their drug discovery and drug optimization process.

The role of computational drug design is to aid in the discovery and optimization of new candidate drug molecules. The drug discovery cycle (Fig. 4.1) can be split into six phases: discovery and lead generation (typically 1–2 years), lead optimization (1–2 years), *in vitro* and *in vivo* assays (1–2 years), toxicology trials (1–3 years), human safety trials (1 year), and human efficacy trials (1–2 years). This analysis leads to a total development time of between 6 and 12 years and development costs of $100 to $200 million or more. It is the express goal of computational drug design to significantly improve this cycle. The remainder of this chapter discusses various methods of computer-based methods for engineering new drugs, be they proteins, peptides, oligonucleotides, or small (less than 700 mw) organic molecules.

4.2. PROTEIN AND ANTIBODY ENGINEERING

The first therapeutically successful products from biotechnology were all proteins (see the Introduction), and many therapeutic proteins have found their way from the laboratory to the clinic. Reliable expression of large quantities of a pure therapeutic protein in a foreign host such as a bacterium or yeast is almost always complicated by the fact that the cellular machinery required for post-translational modifications is not readily available. Typically changes need to be engineered in the protein in order to allow it to be expressed and isolated in a folded, fully functional form. The goal of protein engineering is to mutate the structure of a protein in a predictable manner so as to generate a protein molecule of chosen affinity or activity. In particular, therapeutic protein engineering seeks to generate protein molecules that can have therapeutic effect either directly, as in proteins required for the amelioration of a disease, or indirectly by acting as inhibitors or antagonists to other molecules, or as carrier molecules for ligands that can then exert an antagonistic or cell-killing effect (Blow et al. 1986; Blundell et al. 1989). Initial at-

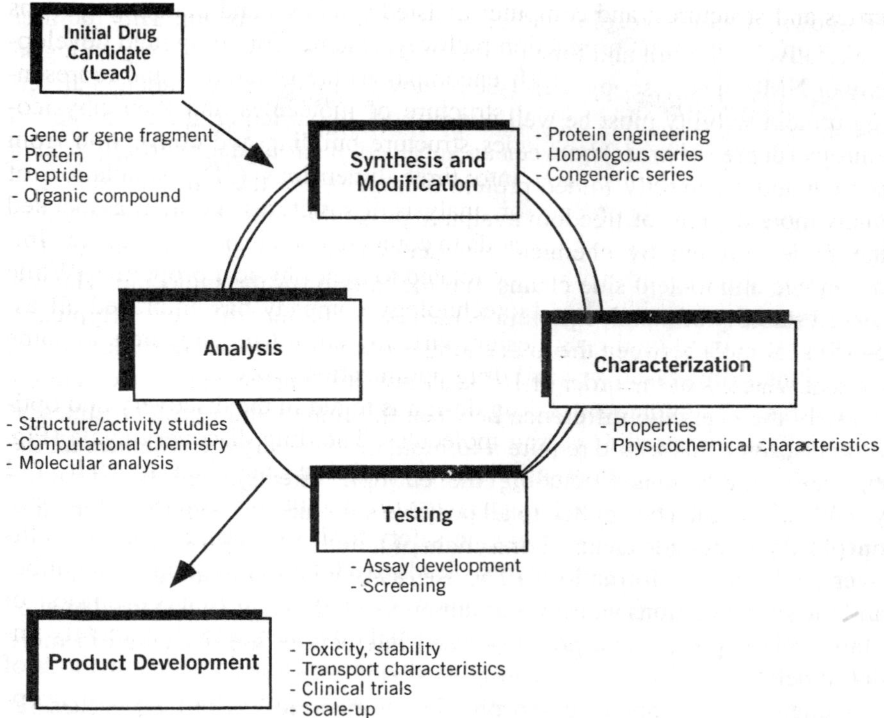

Figure 4.1. The drug discovery cycle.

tempts at engineering changes into proteins resulted in a surprising range of effects caused by single mutations when only one was expected; the results of these experiments reveal how little was then known about the principles of protein stability and the energetics of ligand binding and catalytic efficiency. These studies also hint at the difficulty of designing a stable protein *de novo* with a prescribed function. Nonetheless, a considerable amount of effort has led to some notable successes in this field, such as engineering stability into proteins by adding novel disulfide-bond forming cysteine residues, increasing the catalytic efficiency of enzymes by selective modification of residues at their active sites, and using engineered monoclonal antibodies as therapeutics and catalytics.

Protein engineering requires as its starting point a thorough understanding of protein structure, function, and stability. For instance, the difference in energy between the transition states of a functional enzyme and a functionless point mutant can be less than that contained in a single hydrogen bond (e.g., as found in the enzyme rubisco; ribulose-1,5-bisphosphate carboxylase/oxygenase). Second, the precise three-dimensional structure of the protein must be known at atomic resolution before engineering can begin. This requires the relatively difficult and time consuming techniques of X-ray crystallography or NMR spectroscopy (described below). Finally, the parameters affecting protein stability must be well understood. In particular, the difference in entropy (degree of order) between an unfolded, misfolded, or partially folded protein and a correctly folded protein is large (since the unfolded state has many more degrees of freedom available to it); this must be compensated in the folded protein by chemical interactions such as the burying of hydrophobic amino-acid side chains into the interior of the molecule. The observed stability of a native protein is the result of a tiny difference (typically 5–15) kcal/mol) between the total energies of the folded and unfolded states, each of which is of the order of 10^6 kcal/mol. It is impossible to calculate accurately the energetic difference between the folded and unfolded states of proteins, since this would require a formula for the energy that defines all the energetic interactions (including the entropic effects), and this formula would have to then be applied to all possible structures of both the folded and unfolded polypeptide chain. (For a chain of length 100 amino acids, there are over 20^{100} such conformations.) The use of energetic force-field calculations and the simplifications and some assumptions within them when used to simulate protein–protein and protein–ligand interactions are discussed in more detail below.

Nonetheless, computer-based protein engineering has proved successful in designing proteins with greater stability, better binding energies, novel catalytic mechanisms, and even the *de novo* construction of stable proteins with prerequisite functions. Of particular relevance is the field of antibody engineering, wherein these ubiquitous therapeutic molecules can now be created with virtually any desired specificity and subsequently engineered for diagnostic, imaging, or therapeutic use thanks to the technologies of monoclonal hybridoma and phage display production. Antibody engineering is described below.

Knowledge-Based Protein Modeling

The speed with which novel protein targets may be cloned, isolated, expressed, and sequenced compared to the relatively slow time it takes to deduce their atomic resolution structures has led to a number of algorithmic techniques to deduce three-dimensional (3D) structure from linear sequence. Theoretically a protein 3D structure should be deducible from first principles (*ab initio*); however, in practice, even a short peptide or protein can, in principle, adopt so many conformations that systematic generation and evaluation of all of them by computer is virtually impossible. Other *de novo* prediction methods typically start with secondary structure predictions, alignments of a number of similar sequences, or analysis of hydrophobic or other patterns such as evolutionary criteria. In general these methods have at best a 60–70% accuracy, and the route from secondary structure elements to their spatial assembly into a complete tertiary structure is currently virtually impossible. A more pragmatic approach uses knowledge of the three-dimensional structures of *related* proteins (assuming they can be found) to direct the orienting of the sequence into a complete three-dimensional fold.

Nearly 50% of newly solved experimental protein structures appear to be related to known folding motifs, consistent with the idea that only a limited number of folds, perhaps as few as 1000, have been generated by evolution. Even if a database of every fold can be established, for knowledge-based modeling to be valuable in therapeutic drug design, specific differences need to be noted in comparing single structures to "consensus" folds such as the positioning of the amino-acid side chains that are critical for enzymatic or regulatory function. Comparative model building, first attempted on enzymes with known similar catalytic function, has now been generalized by Greer and by Blundell and colleagues. The procedure requires the careful superposition of known homologous crystal structures in order to identify structurally conserved regions (SCRs), and to be able to distinguish these from structurally variable regions (SVRs), usually loops at the protein surface. The sequence to be modeled is then added to the structure-based sequence alignment, constraining amino acid insertions or deletions to loop regions and not (usually) within the SCRs. Loops are built by a combination of procedures, including systematic conformational search, protein loop database searching, and homology comparison. Amino-acid side chains can be positioned by superposition of comparable atoms, analysis of "rotamer libraries" (catalogs of the statistical preferences for side-chain positions in different environments obtained from known structures), and/or conformational search and energy minimization (Blundell 1994). Model structures of therapeutic interest that have been built in this way include renin, C5a inflammatory protein, and HIV protease. More recently the analysis has been extended to include the surface domains of membrane receptors such as the IL-2 receptor, P-selectin, and gp39, and the G-protein-coupled receptors. Because of the strong structural conservation of antibody molecules, except in their hypervariable loops (the so-called complementary determining regions or CDRs; see Fig. 3.2), these molecules are particularly amenable to modeling by a combination of homology, knowledge-base, and constrained conformational search procedures (Rees et al. 1994). The therapeutic value that can be derived from accurate three-dimensional structures of antibodies is discussed below.

Monoclonal Antibody (MAb) Engineering for Immunotherapy

Ever since Erlich's visionary proposal nearly a hundred years ago to use antibodies as "magic bullets" that target and eliminate diseased cells, scientists have longed to use these natural, highly specific molecules in much the same way as the human immune system—to combat disease. The delineation of antibody gene and protein structure during the 1980s together with the advent of hybridoma technology now allows the expression of antibodies from a single clonal lineage (monoclonal antibodies; see Section 5.1 and Fig. 4.7 for more on the structure of these molecules). These results have led to the first experiments in antibody therapy. The principle behind antibody engineering therapy relies on the ability of the antibody to inhibit a specific target that impacts a fundamental mechanism such as blocking a pathological enzymatic cascade sequence. In addition, for antibody therapy to be successful, one must avoid immunizing the patient with the antibody. This problem arises because therapeutic monoclonal antibodies can only be generated in quantity in mouse cells. The immune response to the therapeutic murine MAb is commonly termed the human anti-murine antibody (HAMA) response, and it has been widely reported with therapeutic or imaging (diagnostic) murine MAbs. HAMA responses have been reported against constant domains, variable regions, and idiotypic determinants. The replacement of murine domains with human domains in the MAb protein would be expected to reduce immunogenicity, although the incidence of immune response seems to vary widely with each antibody. Experience with chimeric mouse/human monoclonal antibodies with a focus on the antibody response to the CDR regions has led to an animal model that may aid in pre-clinical evaluation of the therapeutic recombinant antibodies with regard to their V-region immunogenicity. Predictions of the human immune response to CDR-grafted antibodies in clinical trials may help validate this model. (Antibody CDR structure is illustrated in Fig. 4.2 on pages 116–117.)

EXAMPLES OF THERAPEUTIC AND DIAGNOSTIC HUMANIZED ANTIBODIES

Antigen-Binding Determinants and Anti-proliferative Activity in Humanized Anti-P185her2 Antibody (Genentech, Inc.)

Genentech has studied the antigen binding of a humanized antibody and the stability and chain association of the various antibody. p185her2 (human epidermal growth factor receptor 2) is a transmembrane tyrosine kinase homologous to the epidermal growth factor (EGF) receptor. It is overexpressed in 30% of breast and ovarian cancers, and its levels of amplification has been linked to poor prognosis of these conditions. A murine MAb (4D5) has recently been raised against the extracellular domain of p185her2, but does not bind to EGF receptor and thus inhibits tumors that overexpress p185her2. The approach to humanizing this antibody has been to use computer modeling of the X-ray structure coupled with sequence analysis to alter those residues that may contribute to a HAMA response. The strategy is:

- choose a consensus human framework from the V_L-kappa and V_H-III genes (see Fig. 4.2 and Chapter 5 for details regarding the naming conventions).
- build a model to identify key framework residues.
- determine the 3D structure by X-ray crystallography to confirm the interactions between framework and CDR residues.

The X-ray model can also be used to identify solvent accessible non-CDR segments that are different between the humanized and murine variable domains. Those segments are then replaced with the corresponding murine sequence and the resulting chimeric MAb tested for its antiproliferative activity. The functional epitope of the hu4D5 molecule was determined by Alanine-scan mutagenesis of the hu4D5 CDR loops, and their effect on the disociation constant K_d measured by kinetic of binding to immobilized p185her2-ECD using the Pharmacia Biacore system. Enthalpies and energies of binding (Δ-H and Δ-G effects) were measured from calorimetry, and these data were used to guide further engineering of the CDR surface. Four residues (His-91, Arg-50, Val-95, and Tyr-100) were found to dominate the free energy of binding. These residues were found to make a shallow pocket on the surface of the antibody. Many other residues contact the antigen but do not make a favorable contribution to the entropy (Δ-G) because of enthalpy–entropy compensation. Tyrosine residues on the Ab surface also contribute to binding via the hydrophobic effect. However, side-chain hydrogen bonds do not appear to provide a large driving force for antigen binding.

X-ray studies of antibody/antigen complexes suggest that the framework residues that support the CDRs also contribute to CDR conformation and antigen binding. These residues vary in antibodies and also contribute to the domain–domain interactions maintaining the association between the light and heavy chains. Additionally the first few amino acids at the N-termini of the light and heavy chains are also seen to be interacting with the CDRs. The X-ray studies have shown that the combining site is formed primarily by the CDRs, although the framework residues also

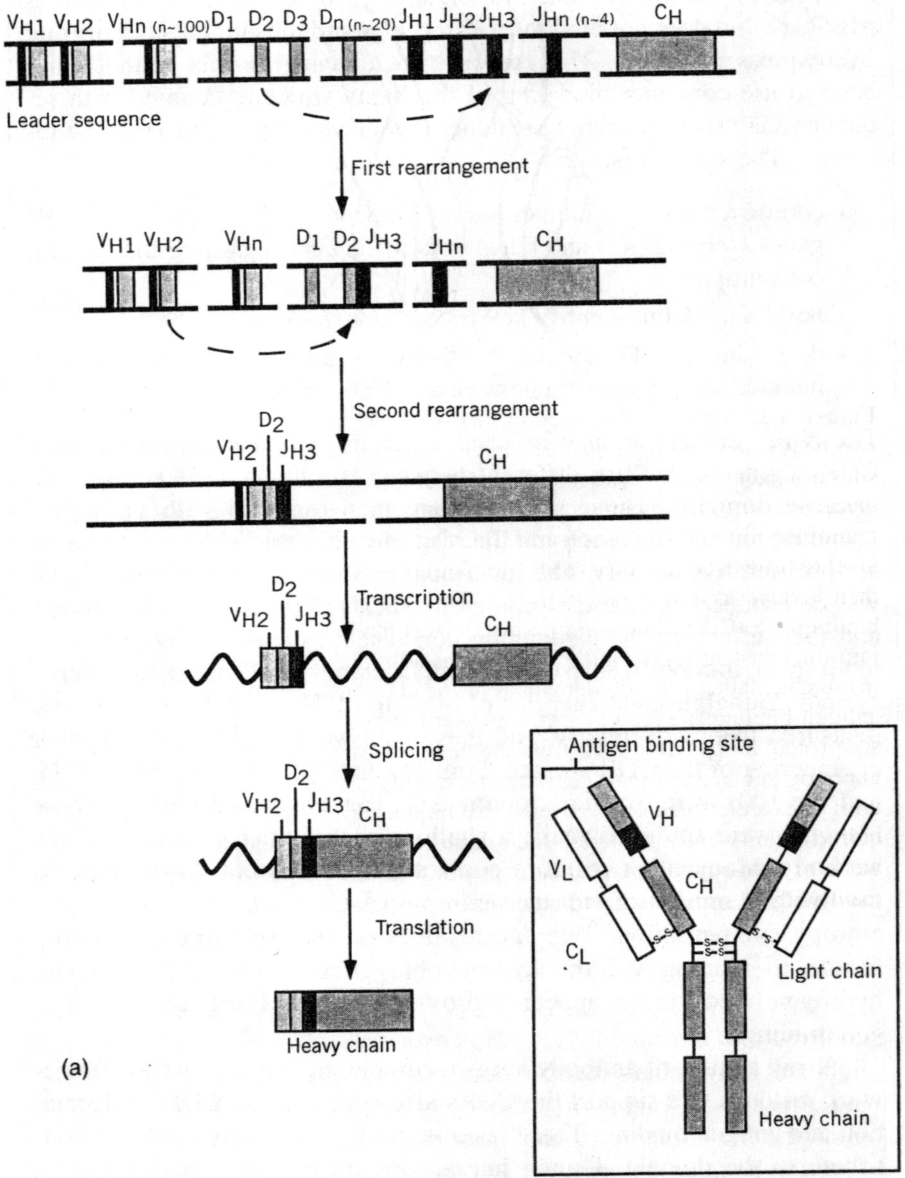

VH1 VH2 VHn (n~100) D1 D2 D3 Dn (n~20) JH1 JH2 JH3 JHn (n~4) CH
Leader sequence
First rearrangement
VH1 VH2 VHn D1 D2 JH3 JHn CH
Second rearrangement
D2
VH2 JH3
CH
Transcription
D2
VH2 JH3
CH
Splicing
D2
VH2 JH3 CH
Translation
Heavy chain
(a)
Antigen binding site
VH
VL
CH
CL
Light chain
Heavy chain

Figure 4.2

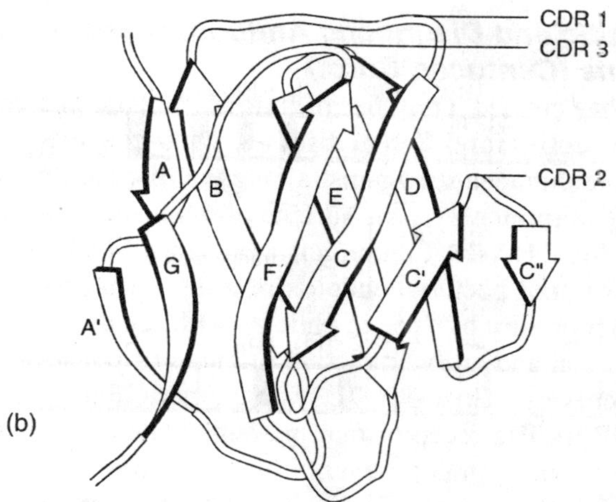

Figure 4.2. Antibody engineering. (*a*) A typical antibody molecule composed of two heavy and two light chains. The interchain linkages and two identical binding sites are indicated: V—variable; C—constant; H—heavy; L—light. *Heavy chain rearrangement:* The upper section represents the genomic organization of the immunoglobulin heavy chain locus. The first rearrangement occurs at the genomic level to bring a D (diversity) and J (joining) region together. A second rearrangement then juxtaposes a V region to the DJ. This V_{DJ} will encode for part of the antigen binding site (CDRs in panel *b*). This DNA is then transcribed and processed to adjoin the constant region to the VDJ. Translation then yields a heavy chain product. (*b*) A schematic of the F_V domain at the head of an antibody molecule showing the antigen-binding site (complementary determining region or CDR) constructed from six protein loops, three each from the heavy chain and the light chain. The goal of antibody engineering is to generate and modify the CDR loop amino acid residues until ones are created or discovered that bind a particular therapuetic target with required affinity and specificity; then these loops must be engineered onto a framework of beta-strands (A-G) that supports their unique three-dimensional arrangement without stimulating an immunogenic response.

contribute to the binding site. However, the mode of association of the light and heavy chain variable domains often changes upon antigen binding. Since these vary from antibody to antibody, it is important to determine the 3D structure of each new molecule. In general, the CDRs are found to be quite deformable, and their structures are influenced by their surroundings. Despite this, p185her2 Ab has proved extremely valuable in the clinic and is currently completing clinical trials and introduction to the marketplace.

Therapeutic and Diagnostic Antibodies and Antibody Fragments (Centocor, Corp.)

Centocor has created a number of murine, human, and chimeric antibodies as therapeutics for different diseases. Their goal is to use these proteins as injectable therapeutics against a range of diseases by targeting specific antigenic determinants on the surface of molecules that represent the disease pathology. HA-1A Centoxin is a human IgM antibody specific to the lipid A portion of bacterial endotoxin. It has been shown to provide a reduction in mortality by 36% in patients with sepsis. (Sepsis is due to bacterial infection and an overreactive immune response stimulated by bacterial endotoxins.) CentoRX 7E3 is a chimeric Fab IgG-1 specific to the platelet GP IIb/IIIa receptor that prevents platelet aggregation following stroke, with no thrombotic events (production of clots) in high-risk patients. The Centara (anti-CD4) chimeric IgG-1 antibody is specific to CD4+ T cells and depletes their presence in patients with rheumatoid arthritis (RA), multiple sclerosis (MS), Crohn's disease, and other autoimmune diseases. In RA patients a 50% reduction in painful joint occurrences is observed. Centocor's conclusions from clinical trials with these antibodies suggest that chimeric antibodies are less immunogenic then their murine counterparts and that small doses of murine Fab imaging agents do not appear to be immunogenic, suggesting that most types of antibodies have low toxicity profiles. Specificity *in vivo* may be complicated by a number of factors, such as the general hydrophobic attraction of the CDR surface, possibly explaining the spectacular failure of monoclonal antibodies for sepsis from Centocor, Synergen, and others. However, Centocor's ReoPro chimeric antibody has recently received FDA approval as a therapy for certain cancers, which suggests a promising future for these molecules as therapeutic agents in specialized cases.

Antibodies in Rational Drug Design (Immunopharmaceutics, Inc.)

Rational drug design is best accomplished when the 3D atomic structure of the target is known and can serve as a starting point (template) for design studies. In the absence of this information, some companies are using the power of the immune system to access the 3D information contained in pharmacological receptors, enzymes, and viruses without the need for special purification and isolation effort of these molecules.

Through recombinant monoclonal technology, antibodies to key drug interaction sites in receptors, enzymes, and viruses can be raised and selected in any desired amount without having isolated the target (these antibodies "mirror" their target sites and thus are labeled *idiotypic antibodies*). The idiotypic antibody gains its specificity from the selection of residues on the surface of the CDR loops, which form a complementary surface to the drug target site. This idiotypic antibody is ideally a mirror image of the target site, in both a spatial and an electronic sense. By using the idiotypic antibody as an antigen, an anti-idiotypic antibody can be produced that is a positive image of the target site. Extraction of the structure of this molecule should then depict the spatial and electronic distribution of the CDR surface that contribute to the hydrophobic, electrostatic, and hydrogen bonding effects that together account for the affinity of the antibody to the target. In addition to X-ray crystallography, other techniques may also be used to "read" out the structural information from this pharmacophore, though not in as much detail. For instance, PCR allows the sequence of the CDRs to be determined. Key residues can then be identified by site-directed mutagenesis and their effect on antibody affinity determined. (The determination of a key Arginine-Tyrosine-Glutamine or RYD sequence in the antibody to fibrinogen receptor was accomplished in this way.) Evidence for the location of key amino acid sequences in the hypervariable loops of anti-idiotypic antibodies can also be determined from their homology to sequences in the initial antigen.

Use of Molecular Recognition Units (MRUs) as Radioimaging Diagnostics (Cytogen, Corp.)

MRUs are small peptides derived from MAbs that have the same binding specificity as the parent MAb. PAC1.1 is an IgM antibody that is specific for activated platelets and that binds to the fibrinogen receptor. The most variable portion of this molecule is CDR H3, which has an exceptionally long insert and contains an RGD (Arg-Gly-Asp) sequence that is thought to provide optimal binding. A tandem repeat sequence of this motif is considered the principal component of specificity and affinity. A peptide containing this motif has been fused to a peptide chelator sequence (KCTCCA), derived from metallothionein, that chelates the radioactive isotope Technium-99 (99-Tc) to produce a bi-domain peptide with a recognition unit (the tandem RGD repeat), and an effector unit (the imaging 99-Tc). Studies in animal models (rabbit and dog) show that the molecule gives specific images of platelet aggregation in 15 minutes to 1 hour.

Genetically Engineered Antibodies for Cancer Diagnosis and Therapy (NCI/NIH)

A monoclonal antibody B72.3 has been shown to react with the tumor-associated antigen 72 (TAG-72). This molecule is a large molecular weight mucin expressed on the surface of a variety of carcinomas including gastrointestinal, mammary, ovarian, prostate, and (non-small-cell) lung carcinomas. Clinical trials of B72.3 for diagnostic tumor targeting in over 1000 carcinoma patients show localization in about 70–80% of carcinomas.

A second-generation rodent anti-TAG antibody was then developed with a higher antigen-binding affinity and more efficient targeting of human colon carcinoma xenografts. Preliminary phase I trials of a 131-idoine-labeled MAb (CC49) suggest that it is a potentially useful clinical reagent in targeting human colorectal carcinoma lesions. However, the potent immunogenecity of these rodent MAbs is an impediment to their clinical application in humans. A recombinant chimeric MAb containing the human gamma-1 constant region has been developed that localizes at similar levels to human carcinoma xenografts. Site-directed mutagenesis has allowed development of an aglycosylated MAb, cB72.3. The aglycosylated molecule is cleared from the peritoneal cavity faster than the parent chimeric MAb. Using PCR technology, truncated versions of the cMAb 72.3 heavy chain have been produced, and studies in athymic mice (which succumb more readily than humans) show that these CH2-domain deleted MAbs localize to tumors and are cleared from the blood faster than the parent MAb. A single-chain variable domain Ab (scF_v) derived from CC49 can also be radio labeled with 125-I or 131-I. This molecule still binds efficiently to TAG-72 expressing tumors, although at an eight-fold lower affinity.

4.3. PEPTIDE AND PEPTIDOMIMETIC ENGINEERING

An advance in therapeutic protein design is the ability to reduce these complex macromolecules into small functional units that are amenable to high-resolution structure analysis and rapid modification. The dissection of multidomain proteins into small, synthetic, conformationally restricted components is an important step in the design of low molecular weight nonpeptides that mimic the activity of the native protein. The intention is to design mimetics of functional domains that retain the same specificity and therapeutic potential of the full protein, and can also serve as templates for small molecule drug design.

In solution, peptides are highly flexible molecules whose structure is strongly influenced by their environment (Marshall et al. 1979). Their essentially random conformation in solution complicates analysis of their receptor-bound or bioactive conformations. The surface location of turns on proteins has led to the recognition that turns play an important role in a number of molecular recognition events (Rose 1988). Mimetic of so-called protein reverse turns have become powerful tools for studying receptor recognition. Historically, Hughes et al. (1975) were among the first to recognize that the linear pentapeptide enkephalin and the condensed heterocyclic morphine both bind to the opiate receptor and elicit analgesia. The ability of a small, organic molecule to mimic the effect of a naturally occurring peptide bode well for the design of peptidomimetic drugs. Good mimetic design then requires knowledge of the ensemble of pharmacophoric information present in the primary sequence of the peptide as well as and the three-dimensional presentation of this information.

Peptidomimetic design has been attempted by screening of natural products for lead molecules (Goetz et al. 1986; Evans et al. 1986), by structure elucidation of the receptor-bound conformation (de Vos et al. 1992), and by application of conformational constraints to search for rigid analogs (Motoc et al. 1986). One approach is to use the known secondary structure elements of proteins (alpha helixes, beta strands, reverse turns) as templates for mimetic design. Reverse-turn mimetics must be able to accurately resemble the diversity of turns that have been classified in proteins. All reverse turns consist of four amino acids constrained by the distances between the alpha-carbons of each of the four amino acids. Design of a mimetic requires the synthesis of the appropriately positioned functional groups (in 3D space) on a relatively rigid framework—a far from trivial problem requiring the stereo- and enantio-controlled introduction of a minimum of four noncontiguous assymtric centers. Nonetheless, the importance of reverse turns in therapeutic recognition events has led to much study of their structure. The retrosynthetic synthesis of reverse-turn peptidomimetics is illustrated schematically in Fig. 4.3 (A–E on page 122).

One class of peptides to which it is therapeutically valuable to design mimetics are the reverse-turn loops found on the surfaces of many molecules and receptors. The cloning and expression of members of the immunoglobulin (Ig) superfamily, including antibodies, T-cell receptors, CD4 and the other CD receptors, and cell adhesion molecules (CAMs), has led to the development of a series of novel therapeutics that utilize these molecules or their antagonists to regulate disease states in a highly specific manner. Each of these molecules plays a role in the mediation of cell surface recognition by utilizing six reverse-turn loops on their surface, the so-called hypervariable complementary determining regions (CDRs). Nonpeptide mimetics to these CDR loops, such as the anti-idiotypic monoconal antibody 87.92.6 which binds to the cellular receptor of type-3 reovirus, are now under development (Saragovi et al. 1991).

Reverse-turn mimetics to Ig superfamily CDRs have been designed using solid-phase peptide synthesis protocols commencing with a peptide chain to which modular components are retrosynthetically attached (Nakanshi et al.

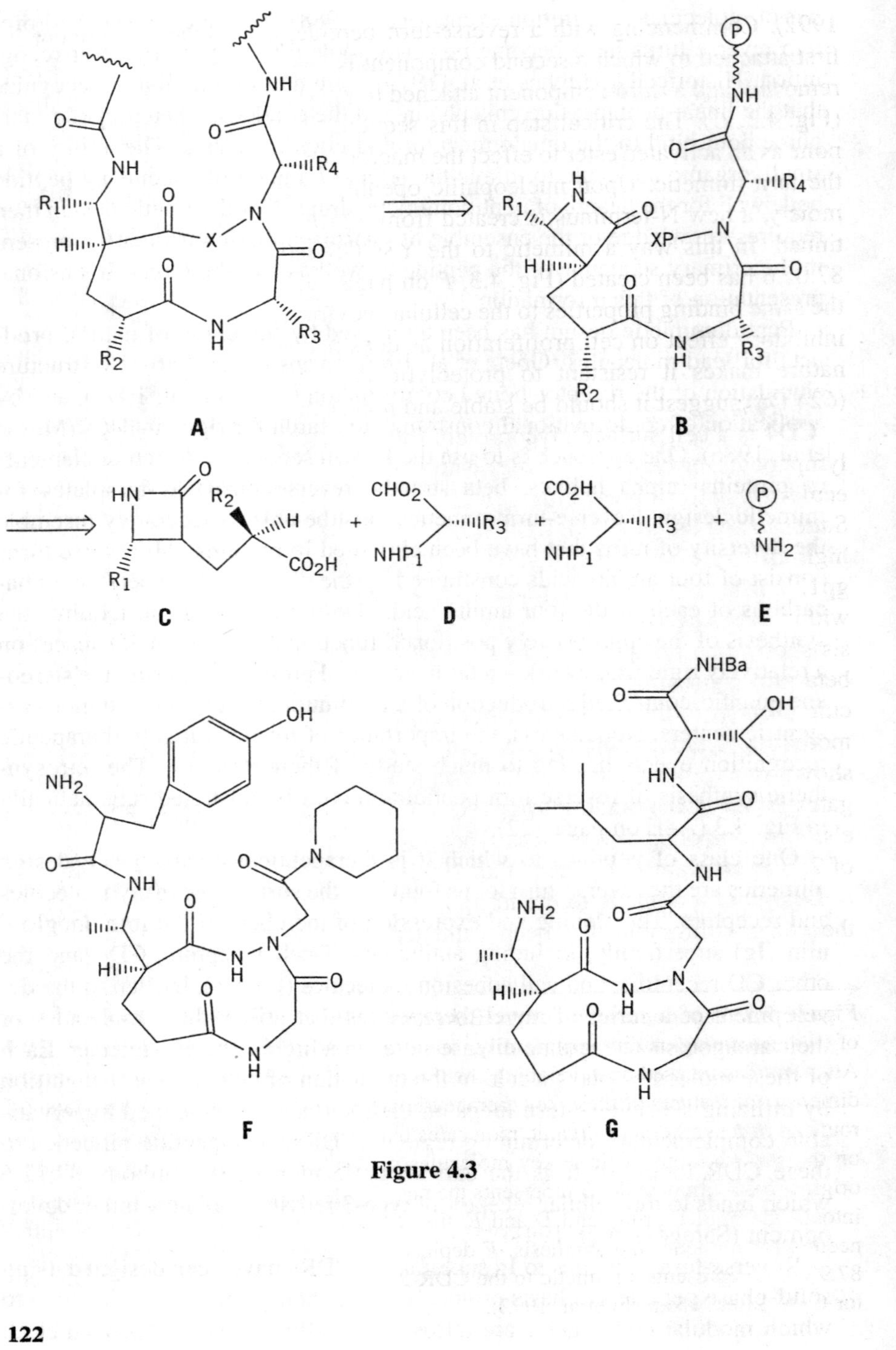

A B

C D E

F G

Figure 4.3

1992). Commencing with a reverse-turn peptide, a modular component is first attached to which a second component is coupled, the protecting group removed, and a third component attached to provide the nascent reverse turn (Fig. 4.3, *C*). The critical step in this sequence involves the use of azetidinone as an activated ester to effect the macrocyclization reaction and provide the turn mimetic. Upon nucleophilic opening of the azetidinone by the X moiety, a new N-terminus is created from which the synthesis may be continued. In this way a mimetic to the YSGGS sequence in CDR2 of Mab 87.02.6 has been created (Fig. 4.3, *F* on page 122). The compound exhibits the same binding properties to the cellular reovirus receptor and has the same inhibitory effect on cell proliferation as does native Mab 87.92.6. Its cyclic nature makes it resistant to proteolytic degradation, and its small size (624 Da) suggest it should be stable and nonimmunogenic.

CD4 is a cell surface glycoprotein found principally on the surface of T lymphocytes, where it associates with class II major histocompatibility molecules on antigen-presenting cells (Gay et al. 1987; Sleckman et al. 1987; Sattenau and Weiss 1988). CD4 also serves as an attachment site for HIV via high-affinity binding ($K_d \sim 10^{-9}$ M) to the HIV envelope glycoprotein gp120. Mutagenesis and peptide mapping data show that amino acids 40–55 within the CDR2-like loop of CD4 is critical for gp120 binding. X-ray analysis shows that residues Gln-40 through Phe-43 reside on a highly exposed beta-turn connecting the C′ and C″ atoms of the beta strands. A small molecule mimetic to this region has been shown by NMR analysis and molecular modeling to closely mimic the conformation of the loop. The compound, shown in Fig. 4.3 *G* on page 122, has a molecular weight of 810 and abrogates HIV-I (IIIB) binding to gp120 in CD4+ cells at low micromolar levels. The compound also reduces syncytium formation 50% at a concentration of 250 micrograms/ml (Chen et al. 1992).

Design of a mimetic to the fibrinopeptide A7-16 (FPA), the substrate for the enzyme thrombin, has been validated by studying the structure of the

Figure 4.3. Peptidomimetic design. Beta-turn loops are a key feature of members of the immunoglobulin superfamily, such as antibodies and T-cell receptors. Attention has focused on developing small molecule mimetics that mimic the three-dimensional features of these key therapeutic proteins. The figure demonstrates one route of retrosynthesis of organic molecules that are mimetics to beta-turn peptides on the surface loops of these key molecules of the immune system. *A* depicts the original beta-turn peptide. *B* represents the nascent mimetic that can be decomposed into *C*, the mimetic ring, and *D* and *E*, the second and third modular components needed for mimetic ring synthesis. *F* depicts a mimetic to the V_L CDR-2 of mAb 87.92.6. *G* represents a mimetic to the CDR-2 like region of the cell surface receptor CD4 (after Nakanishi et al. 1992).

thrombin-FPA complex by X-ray crystallography. Thrombin plays a critical role in the clotting process, effecting the removal of fibrinopeptides A and B of fibrinogen by the selective cleavage of two Arg-Gly bonds selected from the 181 Arg/Lys-Xaa bonds in the intact fibrinogen molecule. Controlled and selective interference of this enzyme is made difficult by the high degree of sequence similarity within the trypsinlike family of proteases (of which thrombin is a member). These enzymes typically contain a trypsinlike pocket with specific insertions that modify their specificity by interacting with additional substrate components (Magnusson et al., 1976). A model for the bound FPA substrate and the enzyme has been determined by a combination of NMR analysis, computer-assisted molecular modeling, and the synthesis and study of peptidomimetic subtrates and inhibitors (Nakanishi et al. 1992). The X-ray structure of a chloromethyl ketone derivative of FPA (FPAM) shows that the amide angles (psi, phi) in the critical P1-P2-P3 pocket of the enzyme and the model are similar, and that they are also similar to the structure of the D-Phe-Pro-Arg (DPACK)-thrombin structure (Wu et al. 1993). One result from the thrombin study is that utilization of a canonical loop motif of a natural peptide inhibitor as a lead for designing reverse turn peptidomimetic inhibitors may provide a general strategy for introducing specificity into an inhibitor.

4.4. COMPUTATIONAL CHEMISTRY IN SMALL-MOLECULE DRUG DESIGN

Although proteins, antibodies, and peptides have had tremendous individual successes as therapeutic molecules (Lilly's recombinant insulin, Amgen's erythropoietin, Schering-Ploughs's interferon-beta), the consensus emerging among biotechnology companies is that small organic molecules, including those that may structurally mimic the role of peptides and proteins, stand a far better chance of becoming successful pharmaceuticals. In part this is because proteins and peptides are difficult to deliver to their target cells, cannot be ingested orally, and are typically metastable, and in part because the pharmacological properties of small organic molecules are far better understood. The advent of high-throughput (mass) screening and molecular diversity technqiues (see the previous chapter) now provide small biotechnology companies the capability to generate and test novel compounds as rapidly as the large pharmaceuticals. Nonetheless, none of these methods are guaranteed to provide a company with an ideal therapeutic molecule, only lead compounds that must be optimized to become therapeutic initial new drug (IND) candidates. Traditionally optimization has required the experience and intuition of medicinal and synthetic chemists. In an effort to rationalize this process of optimization, biopharmaceutical companies have turned to computational chemistry as a method of providing intelligent heuristics (rules) for drug design. The role of the computational chemist is to perform quantitative and qualitative modeling of candidate chemical structures in order to elucidate and improve upon their pharmacodynamic properties and, in so doing, to develop heuristics that can be repetitively applied to the efficient discovery of new drugs.

Designing Drugs without a Target 3D Structure

If the precise three-dimensional structure of the target protein is unknown, small molecule drugs may still be found or built against it, but considerable experimental data are required to guide the design process. In particular, it is imperative that a diverse series of compounds are isolated that are all known to be active at the target (Dean 1987). Using a combination of quantum and classical chemical techniques, chemical compound database searching, and statistical and combinatorial methods for mapping out the properties of compounds into a "function space," it is possible to identify the appropriate backbone atoms and side-chain functional groups that need to be modified in order to systematically improve potency and reduce toxicity. Quantum mechanical methods allow the electronic distribution on atoms and functional groups to be determined (this information is important in understanding the bonding characteristics of these moieties). Classical (Newtonian) molecular mechanics allows the simulation of the motion of the atoms in a compound and helps determine its flexibility. This is useful because compounds typically bind to the target in a conformation (three-dimensional atomic arrangement) that is "strained" away from the minimum energy (free) conformation. Knowledge of a series of molecules all known to be active at a receptor site allows one to determine the spatial arrangement of functional atoms or groups that are common and that together account for the observed activity in that group. Functional classes typically include atoms that are hydrogen-bond donors or acceptors, hydrophobic regions such as aromatic rings, and charged or partially charged (polar) groups that can participate in electrostatic interactions with complementary regions in the receptor site. Computer methods can probe a series of compounds automatically in order to determine this arrangement for a set of compounds—the set is then known as the pharmacophore (Martin 1988; Marshall and Cramer 1988; Appelt et al. 1991; Kuntz 1992). Since more than one possible pharmacophore hypothesis can exist for a given set of molecules, systematic pharmacophore mapping the and study of alternative pharmacophores is important in order to delineate the spatial distribution of interactions in the receptor site in the absence of direct structural information. (Formally the pharmacophore hypothesis assumes that a particular spatial distribution or "field" of steric, electrostatic, and hydrogen-bonding potential in the ligand causes recognition at the receptor site. This broader definition of a pharmacophore does not require defined atoms, functional groups, or substituents.) Molecule volume or surface measurements of these fields can be used to assess whether two ligands share similar pharmacophoric regions.

Once a pharmacophore is known, it can be used to probe a chemical library by flexible computer-based 3D searching to find other compounds containing the same pharmacophore, or used as a scaffold to build a chemical library by the methods of combinatorial chemistry discussed in the previous chapter. Compounds found, purchased, or constructed can then be assayed against the target, and the information obtained used to expand the pharmacophore hypothesis. By the iterative process of synthesis, screening, and analysis, a prioritized set of compounds can then be created, yielding new leads and eventually a candidate drug for clinical trials. Since each iteration of this cycle generates more experimental data, it is important to have procedures that store the structure-activity relationships generated for each new lead compound or pharmacophore. Quantitating these data may allow new insights into which direction the synthesis strategy should pursue. The concept of quantitative structure activity relationships (QSARs) as a statistical technique to explain observed activity data has come into use in recent years as an objective procedure for analyzing the potentially huge amounts of data that high-throughput screening can generate. Each of these methodologies is described in the following sections.

Computational Aspects of Small Molecule Design

Quantum Mechanical Considerations

Quantum mechanical (QM) methods are useful when the accurate placement of electrons and molecular orbitals are needed, such as in electrocyclic reaction pathways or in determining the electronic distribution, bulk, and polarization effects of functional groups where HOMO (highest occupied molecular orbital) and LUMO (lowest unoccupied molecular orbital) energies and stabilization are desired properties. Properties that require the quantum approach include transition states, reaction pathways, perturbations, electronic effects, polarization, and charge transfers. A series of quantum mechanics programs have been compiled by the Quantum Chemical Program Exchange (QCPE) at the University of Indiana. These have been expanded to include semiempirical treatments (those that attempt to solve the Schroedinger equation of the system using approximations). These types of calculations have also proved useful in determining the molecular properties of functional groups to be used in molecular similarity calculations—an important property to measure when attempting to optimize the diversity of molecules in libraries such as those created by the methods discussed in the previous chapter.

Classical Molecular Mechanics

Typically, when designing a small molecule drug based on either *de novo* or an existing lead compound, a number of issues must be considered. These include what conformations (three-dimensional arrangement of the compounds atoms) are available, which of these conformations are relevant to the compounds activity, and what chemical substitution pattern should be used to optimize the compounds activity. Typical semiempirical molecular mechanics calculations can determine a conformational preference of a torsion angle or the preferred conformation of a heterocyclic ring system, but sophisticated conformational search procedures may be required to determine other possible conformations of a compound with as few as 5 or 6 rotatable bonds. Even for such a relatively simple system (by contrast, proteins have 500 to 1000 rotatable bonds along the main chain alone) this is a significant computational calculation. To efficiently explore the complex multidimensional energy surface available to a molecule, molecular mechanics calculations assume that specific intramolecular forces determine the family of possible conformations and that the molecular system can be considered a collection of particles held together by harmonic forces.

Another use of these force fields is in the method of molecular dynamics, whereby the chosen force field is applied to the molecule at a preset temperature (or range of temperatures) in order to simulate the motion of the molecule at that temperature. By studying the motions of the atoms in a molecule or compound of interest, chemical properties can be derived based on an averaged ensemble of structures rather than the single one generated by energy minimization. Pharmacological uses of the results of molecular dynamics simulations include determining time-averaged geometries of compounds, determining molecular flexibility, and examining solvation effects. However, even on modern computers, simulations of proteins or protein–ligand interactions (such as drug–receptor interactions) are limited to the nanosecond time scale. A related method, quenched molecular dynamics, runs simulations at a constant, elevated temperature which permits the molecule to cross potential energy barriers, and then periodically extracts conformations during the simulation and minimizes these. A similar technique is simulated annealing, whereby high-temperature dynamics are used to allow the molecule to cross a conformational barrier and then the temperature is lowered to allow the molecule to "settle" into a low-energy conformation. This conformation is then refined by energy minimization. The heating/cooling cycle is repeated until a series of distinct low-energy conformations is generated. This procedure has been successfully used to simulate the mode of binding of a beta-lactam binding protein (DD-petidase) using the structure as the starting point for molecular dynamics simulations (Boyd et al. 1991). Data from the X-ray simulation was shown to be consistent with experimental data quantitating the enzyme–ligand interactions.

A critical issue with the use of molecular mechanics force fields is that to date none have proved totally satisfactory for complexes of both large and small molecules, the predominant interactions of interest to the therapeutic biotechnologist seeking to improve or design new pharmacologically useful compounds. Historically the force field MM2 by Allinger and colleagues has been used extensively for "small" molecule calculations (organic compounds of less than 500 atoms), while the ChARMM, GROMOS VFF/Discover, and AMBER force fields have all been well parameterized for use with proteins and nucleic acids.

Conformational Searching and Pharmacophore Mapping

Conformational searching allows chemists to determine the regions in conformation space available to both a receptor and a ligand and provides a framework for the degree of improvement in affinity and/or removal of unwanted side effects from a chemical structure. One approach to conformational searching is the "active analog" approach, whereby the pharmacologically important features of a molecule are placed in the same orientation as those in an analog of known activity, by systematically searching for all possible overlaps (Marshall et al. 1979; Marshall and Motoc 1986). If all the common alignments for a series of molecules can be determined, the active conformation can be expected to be one of these common conformations. In practice, a conformational search is performed, and the spatial relationships among the pharmacophoric groups is noted. Then an analog is searched, but with its pharmacophoric groups constrained to fall into similar positions as those in the first molecule. A new set of constraints is generated, defined in conformational space as the intersection between the two molecules (the set of constraints is known as a distance map). This procedure is repeated for all molecules among the active analogs to refine the constraints that define an active pharmacophore.

Typically chemists can identify a series of active compounds to a given receptor, even though the precise three-dimensional arrangement of active site points (hydrogen-bond donors and acceptors, regions of hydrophobicity, charge centers) among these compounds may not be apparent. This is because the active conformation of each analog bound to the receptor is rarely known and inferences must be made on the most likely bound conformations. For even a limited series of compounds, the number of combinatorial possibilities of conformations is enormous. Tools are now available that automatically identify pharmacophoric site points and then perform constrained conformational searching of the entire series of analogs in order to identify the three-dimensional arrangement of site points that delineates the pharmacophore (Dammkoehler et al. 1989; Motoc et al. 1986). The fundamental assumption in these algorithms is that a single bioactive conformation is common to each of the ligands. In the case of a known receptor (target) structure, receptor target sites can also be included in the search. The heart of such automatic pharmacophore mapping routines is the identification of a set of elements common to each of the active compounds which have interelement distances falling within some tolerance. Algorithms from graph theory such as clique detection can be used to find all feature elements and feature distances of a given molecule that fall wholly within those of a reference molecule (selected from the set as the one with the fewest possible active conformations). A tolerance factor is needed primarily because no fixed set of conformers is likely to contain exactly the same binding conformation. Alternatively, in the case where the chemical features common to a set of compounds is known but their precise three-dimensional arrangement at the receptor (target) site is unknown, a constrained conformational search is possible across all the active compounds to identify conformations that are common to each. Hopefully only a small set of distinct conformations will be found for a given set of chemical features, site points, or feature classes; these then define the three-dimensional characteristics of the pharmacophore. These algorithms will find many possible pharmacophores unless the lead compounds are carefully chosen to discriminate from the possible solutions. Thus compounds should be prescreened to include the least number of starting points (site points) and the least number of possible conformations (unstrained molecules). Minimizing conformational redundancy and maximizing structural variation among the lead compounds optimizes the capacity of the algorithms to generate the fewest number of pharmacophoric maps.

It should be noted that the term "pharmacophore" can have several meanings; in the descriptions above it is assumed that only active compounds are provided. Ideally, however, one would like to be able to distinguish common properties of strong agonists and antagonists so that the pharmacophoric hypothesis developed can clearly distinguish between the chemcial characteristics needed for recognition from those needed for activation. Pharmacophores can also be distinguished according to the sets of molecular properties used to define them such as conformational flexibility, electronic properties, and whether spatial relations between atoms explicitly or instead their chemical properties are used. Some examples of pharmacophores are illustrated in Fig. 4.4 on page 134.

Quantitative Structure-Activity Relationships (QSARs)

One of the most widespread uses of computational chemistry has been in the quantitative analysis of chemical compounds of possible pharmacological interest. The traditional methods of pharmaceutical compound discovery have included a combination of preferred synthesis approaches (building

APPLICATIONS OF PHARMACOPHORE-BASED DRUG DESIGN

Development of a Candidate Pharmacophore From a Series of Delta-Selective Opiod Cyclic Peptides (Molecular Research Inst., Palo Alto, Ca; Chew et al. 1991)

A series of 12 related analogs were selected to develop a pharmacophore model, 9 known to recognize the receptor with high affinity and 3 with very low affinity (Chew et al. 1991). Systematic conformational search of each of these peptides generated a huge number of conformers, which were subsequently pruned by ignoring all those above a 5 kcal/mol cutoff from the lowest energy conformer. (Using an energy cutoff can be rationalized as eliminating structures that are unattainable at physiological temperatures.) Identification of key regions of the molecules based on SAR studies identified the N-terminus Tyrosine (Tyr) region and a second aromatic ring from Phenylalanine-4(Phe) as crucial moieties (Chew et al. 1991). A test of the model was to find conformers that maintained the spatial relationship of these portions of the molecules in the 9 actives but not in the 3 inactives. Three distance criteria were used: the distance between the center of the Tyr-1 and Phe-4 aromatic rings, the distance between the two rings and the terminal amine Nitrogen atom, and the overall root mean square (rms) deviation of the cyclicized portion of the peptide. Another criteria was the general steric requirement that with the aromatic rings in the appropriate spatial position, the remainder of the structure did not occupy areas of the receptor cavity significantly different from the 9 active molecules. Only one conformer satisfied all these requirements, consistent with the hypothesis that the spatial arrangement of the aromatic rings in Tyr-1 and Phe-4 are critical for receptor antagonism. Graphical tools that allow quick and easy visualization combined with criteria for the overlap of the molecules to the model (e.g., minimization of rms differences) helped in the validation of the models.

Design of Potent and Selective D1 Agnosists (Y.C. Martin, Abbott Labs.)

Two receptors respond to dopamine: the D1 receptor that activates adenylate cyclase and the D2 receptor that inhibits it. Traditionally the only known agonists have been of D1, the benzodiazepenes (of which Valium is the most famous example). In an attempt to design a potent and selective D1 agonist that is not of this class, Martin and collaborators studied the structures of molecules known to be active at this site. A series of D1 and D2 agonists are known, and a 3D pharmacophore was generated wtih both active and inactive molecules. The difficulty in determining the pharmacophore model from these compounds was determining the common conformation of the features (Fig. 4.4*a*,). As the figure shows, the unsubstituted phenyl ring and basic nitrogen contribute to the pharmacophore, but it is also observed that the unsubstituted phenyl ring adds 10–100 fold to the affinity for the receptor. Determining the conformation that best overlaps the unsubstituted aromatic ring requires sophisticated search procedures that map pharmacophoric groups and atoms into their common conformations (Martin et al. 1988). These methods have been validated by superimposing inhibitors to endothiapepsin (a structurally well-characterized serine protease) with the crystal structures of the inhibitor-enzyme complexes. In this way series of potent and selective D1 agonists were selected for use in a CoMFA analysis, (see QSAR section below for an explanation of CoMFA) relating their interaction energies to their potency by partial least squares (PLS). This analysis, coupled with an analysis of the positive and negative steric contours delineating the receptor site, led to the synthesis of about 200 compounds with affinities that were 2 to 3 log units greater than the starting compounds. In particular, the analysis suggested that the addition of a phenyl ring would fill a binding pocket in the D1 receptor not present in the D2 receptor. Several molecules designed with this feature are amenable to chiral synthesis and further optimization by 3D-QSAR.

Qsar Analysis Of An Ace Inhibitor Series (Depriest Et Al. 1993)

Angiotensin II is an octapeptide whose physiological effect is vasoconstriction and resultant clot formation. The final step in the conversion of angiotensin II from its inactive prehormone angiotensinogen is the cleavage of a carboxy-terminal dipeptide (Histidine-Leucine) by the angiotensinogen converting enzyme (ACE). Studies of representative ACE inhibitors such as captopril, enalapril, and lisinopril have delineated three key functional features: a C-terminal carboxylic acid group, and amidic carbonyl with hydrogen bonds to a residue at the active site, and a nucleophilic functional group such as a carboxylate, hydroxamate, phosphonate, or thiolate which is speculated to interact with a zinc ion that catalyzes cleavage of the peptide. The skeletons to which these functional groups are attached are remarkably diverse. A reasonable supposition therefore is that all the different inhibitors affect the ACE enzyme the same way; that is, upon binding, all place these pharmacophores in a common relative geometry. Using the active analog approach, Marshall and coworkers searched for common orientations of a set of four pharmacophoric atoms: a zinc atom oriented to be in the optimal location for coordination with the attacking nucleophile, an amidic carbonyl carbon, an oxygen, and an oxygen in a carboxyl group. A systematic search and graphical plot of interpharmcacophore distances displayed the intersection of distributions for a series of 28 different inhibitor molecules. A single distance map was found, which was found to be consistent with 26 additional structural classes of ACE inhibitors. The alignment geometry derived in this way was compared to the experimentally observed orientations of a number of inhibitors to thermolysin, another zinc-containing metalloprotease for which the structure of a number of enzyme-inhibitor complexes are available. The alignment rule derived for the thermolysin inhibitors was compared against that of their bound orientations in order to validate the distance maps generated. The distance map for the ACE inhibitors was then used to align the entire set for CoMFA analysis to seek a predictively useful 3D-QSAR. CoMFA intends to explain the differences in experimental activities between a related set of molecules in terms of the differences between the steric and electrostatic "fields" exerted by those ligands (see the discussion of CoMFA above). CoMFA analysis of 68 inhibitors yielded a PLS model with eight components and a crossvalidated *r*-squared value of 0.66. The CoMFA model was tested by aligning novel inihibitors using the distance map generated from the initial 28 compounds. A plot of predicted versus actual activity generated a predictive *r*-squared of 0.79—an excellent validation of the CoMFA model. The CoMFA coefficient contour maps were calculated as the scalar product of the steric and electrostatic potentials, and the standard deviation of all values in the CoMFA columns were plotted such that differences in field values were associated with differences in inhibitory potency. Strong electrostatic interactions were predicted for the C-terminal carboxyl as well as the amide proton of the P1′ residue (expected to function as a hydrogen-bond donor). The CoMFA field also predicted both a steric and electrostatic interaction about the P1 residue, namely the pendant phenethyl group of captopril that interacts with the hydrophobic pocket. Enhancement in binding of this residue may be through a cation-pi interaction or a hydrophobic interaction stabilizing the quarternary ammonium group and the electrons of the aromatic rings (Dougherty and Stauffer 1990). The ability to reliably predict the affinity of molecules by CoMFA serves the types of functional groups and substituents to be added determine the types of functional groups and substituents to be added based on their contribution to the calculated CoMFA field.

compounds with functional groups for which synthesis routes are present) or brute force approaches (placing substituents at every position in a lead molecule). QSAR (quantitative structure activity relationships) is an alternative approach to improving the activities of compounds based on a series of molecules that have been made and tested. A classical QSAR attempts to correlate an experimental property such as activity (*A*) with calculated structural parameters *a, b, c,* in an equation of the form

$$\log A = x_1 a + x_2 b + x_3 c + \ldots + \text{constant}.$$

Properties of the molecules that can either be calculated from their structure or experimentally derived can be used to define the mathematical relation that predicts the target value from the parameters. The property relations *a, b, c* may have either linear or quadratic representations. If such a relation is found, the QSAR can be used to predict the activity of a new compound of unknown activity that is a member of the same series. A QSAR may also be used to suggest the changes needed to a lead compound in order to improve activity or reduce crossreactivity or toxicity. The parameters attempt to measure the stability of the drug, its pharmacokinetic (transport) properties, its oral availability, and its likelihood of crossreactivity and/or toxicity. Parameters used in the linear relation equation describe physicochemical features of the functional groups such as molecular weight, octanol/water partition coefficient (a measure of lipophilicity, and thus transport across a membrane), fragment shape descriptors, and other substituent parameters. These are usually calculated as the effect on a parameter of converting a hydrogen atom to the given functional group. Substituent parameters that have been validated as descriptors of drug function include π (effect on hydrophobicity—an important parameter in determining drug transport properties, e.g., ability to cross a membrane), σ (effect on electron density at other aromatic positions that are meta or para to the substituent), and molar refractivity (MR) (molecular propensity to bend light rays—a molecular volume effect). Other substituent parameters inlude the Verloop parameters which define the length of a substitutent along orthogonal axes and the Swain and Lupton F and R values which define the effect of a substituent on electron density separated into its inductive (F) and resonance (R) contributions. Traditional QSAR applications have successfully correlated biological activity to these physicochemical parameters for congeneric series of molecules (Hansch and Leo 1979).

Once a series of parameters appropriate to the compound series of interest has been calculated, a QSAR is obtained by applying standard statistical methods such as multiple regression (MR) or partial least squares (PLS) techniques to the experimental values (if they are continuous), or discrimi-

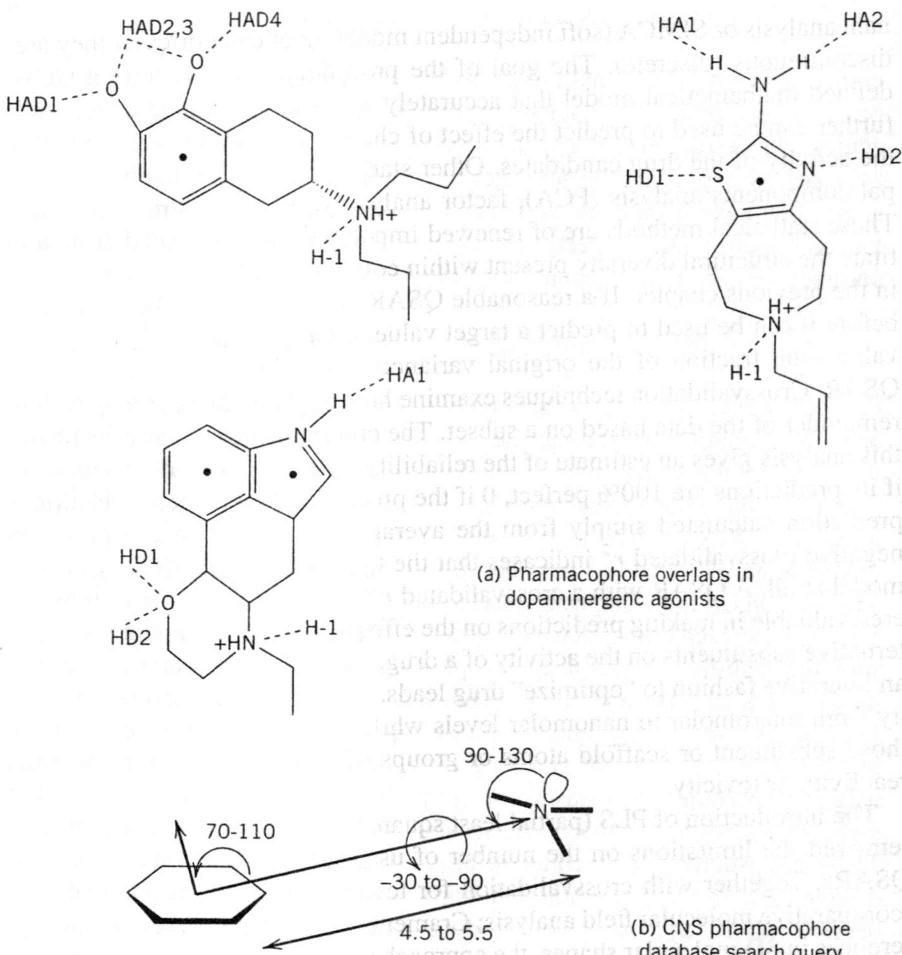

Figure 4.4. Pharmacophore definition. (*a*) Possible points of overlap in three typical dopaminergic agonists. HA refers to hydrogen-bond acceptor sites, HD hydrogen-bond donor sites, H– to anionic hydrogen-bond acceptor sites. Other overlap points in the pharmacophore definition include nitrogen atoms and centers of mass of aromatic rings (after Martin et al. 1988). (*b*) An example pharmacophore for a series of molecules that are CNS receptor antagonists is described. The pharmacophore may be obtained by physical inspection of the molecules or by computer-assisted methods that determine possible three-dimensional superpositions of "features" (atoms, functional groups, aromatic or other ring systems) and other pharmacologically relevant moities (e.g., hydrogen-bond donor/acceptor sites and hydrophobic groups). In this case the pharmacophore consists of a six-membered ring separated by between 4.5 and 5.5 Å to a nitrogen atom with bond and torsion angles as displayed in the figure (adapted from G. Grant and W. G. Richards, *Computational Chemistry,* Oxford Science Publications, Oxford, England, 1995).

nant analysis or SIMCA (soft independent modeling of categories) if they are discontinuous (discrete). The goal of the procedure is to generate a well-defined mathematical model that accurately fits the experimental data, and further can be used to predict the effect of chemical substitution patterns on the activity of the drug candidates. Other statistical methods include principal components analysis (PCA), factor analysis (FA), and cluster analysis. These statistical methods are of renewed importance with the need to quantitate the structural diversity present within combinatorial libraries described in the previous chapter. If a reasonable QSAR is found, it must be validated before it can be used to predict a target value, usually by determining an r^2 value—the fraction of the original variance in the data that is fit by the QSAR. Crossvalidation techniques examine how well the QSAR predicts the remainder of the data based on a subset. The crossvalidated r^2 derived from this analysis gives an estimate of the reliability of the derived QSAR—unity if its predictions are 100% perfect, 0 if the predictions are no better then the prediction calculated simply from the average of all compounds, while a negative crossvalidated r^2 indicases that the QSAR model is worse than no model at all. A QSAR with a crossvalidated r^2 of greater than 0.5 is considered valuable in making predictions on the effect of chemical synthesis of alternative substituents on the activity of a drug. This procedure can be used in an interative fashion to "optimize" drug leads, especially in increasing activity from micromolar to nanomolar levels while simultaneously deleting out those substituent or scaffold atoms or groups that may contribute to cross-reactivity or toxicity.

The introduction of PLS (partial least squares) as a regression method has removed the limitations on the number of usable parameters for detecting QSARs. Together with crossvalidation for testing the QSAR, and CoMFA (comparative molecular field analysis; Cramer et al. 1988) for describing differences in 3D molecular shapes, the approach now provides a powerful predictive methodology to compound design. (The CoMFA methodology is illustrated in Fig. 4.5 on page 136). Improved software interfaces and visualization tools now allow chemists to rationally alter molecules intuitively on-screen and utilize 3D-QSAR by extending the parameters used (Ghose and Crippen 1985; Cramer et al. 1988). CoMFA is a patented 3D-QSAR technique (Cramer and Wold 1991) that assumes that steric, electrostatic, and hydrogen-bonding forces are the major contributors to binding between ligands and their receptors, and that these interactions are shape-dependent. CoMFA requires a series of compounds to be aligned by a known or input pharmacophore and then constructs the steric and electrostatic fields around each of the molecules calculated on a grid by determining the force exerted onto a probe atom placed at each lattice point on the grid. The values of the fields on the grid points are used in a PLS analysis against

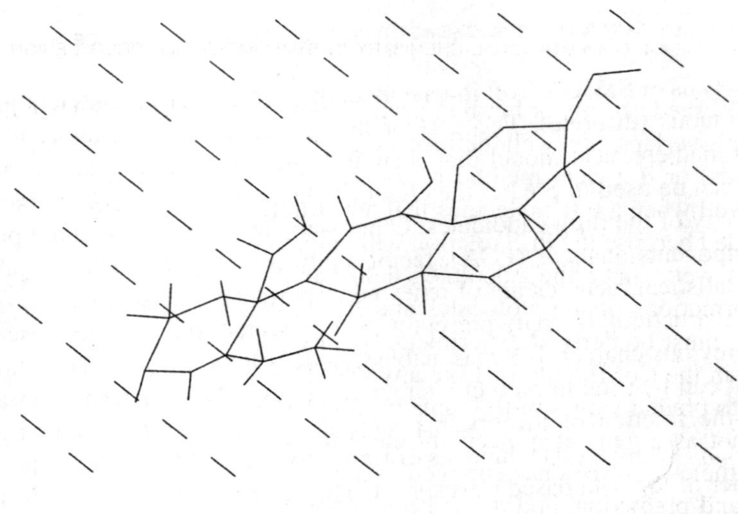

	Bio Act	Steric 1 1000	Elctst 1 1000
CMPD 1	5.1		
CMPD 1	6.8		

PLS

↓

Equation:

$$\text{BioAct} = y + a \times S1 + b \times S2 + \ldots + m \times E1 + n \times E2 + \ldots + z \times E1000$$

Figure 4.5. 3D quantitative structure activity relationships (QSAR) by comparative molecular field analysis (COMFA). 3D QSAR considers the consensus structural elements across a range of different molecules allowing evaluation of the influence of different structural features on the behavior of a drug molecule at the binding site. First, a grid is placed about a collection of superimposed molecules. (It is assumed that the molecules have been aligned in 3D in a manner analogous to their mode of binding to the receptor.) For each molecule interaction energies are calculated at the grid points using a probe atom to calculate steric (van der Waals), electrostatic, and hydrogen-bonding effects. Assuming a simple ten-point cubic grid, for each molecule there are 3×10^3 computed energy values to be correlated to one experimental value—the binding constant. Since the importance of the energies at each grid point will vary with regard to their importance in determining the binding constant, those with small values may be ignored. The 3D-QSAR analysis produces a regression equation with 3000 variables. This vastly underdetermined problem can be resolved by partial least squares, reducing the energies at the grid points to only those clusters that contribute largely to one or other of the components (electrostatic, steric,

the observed activity or binding data. The method works well when the molecular overlaps are well defined (e.g., in planar ring structures such as steroids) or if a clear method for orienting the compounds exists. It works less well when a few molecules that are structurally dissimilar to the rest are included because the PLS relation will not predict the activities of these molecules well. Since the calculated field values are highly dependent on the conformations of the molecules and their relative orientations, "alignment rules" must be derived experimentally or analytically that orient the molecules in the CoMFA study. Like any QSAR procedure, CoMFA can only be used to predict values of the same type as those used to develop the relation, and not as a general molecular determinant of recognition and activation. Nonetheless CoMFA has emerged as a powerful technique in compound design and prediction, and it has been successfully used in a variety of pharmacological areas.

Computational Screening of 3D Databases

A pragmatic approach to finding lead drug molecules is to assemble and then screen large databases of chemical compounds. Significant collections already exist, and it is now possible to purchase large numbers of commercially available compounds. One of the most reliable and widely used databases of organic compounds is the Cambridge database of X-ray crystallographic data, which contains nearly 100,000 organic compound structures. Other databases are available of 2D chemical structures, and reliable tools for generating 3D structures from 2D database-derived connection tables allow these to routinely be converted into 3D for drug design studies (Rusinko and Pearlman 1988; Leach et al. 1990). The techniques of creating combinatorial libraries described in the previous chapter provide a rich source of structurally diverse compounds, and vendors such as Arqule, Pharmacopea, Panlabs, and CombiChem now provide custom combinatorial libraries. Computational screening of these or other compound databases allows one to take a lead pharmacophore or compound and probe the database for other molecules that are structurally similar to the query. In the case where the receptor structure is known, this concept can be extended to allow the definition of included and excluded volumes from the receptor site—the

hydrogen-bonding). An indication of the most important regions of the molecules in determining activity is obtained by graphical color-coded depictions of these clusters. Future molecules may then be designed in such a way as to optimize their properties to these regions in space (e.g., adding steric bulk with bulky functional groups) and their activity predicted using the regression curve (after Cramer et al. 1988).

assumption being that good inhibitors to the target possess significant structural complementarity to their receptor. Binding mode models can be developed using QSARs that are in turn used as constraints in the search for new leads. Typically these models contain the active moieties or substituents determined to be responsible for activity, and their spatial characteristics relative to one another. New techniques that assist in finding compounds in structural databases include:

1. Creation of good pharmacophoric models that define the constraints in computer-based screening by automatic algorithms.
2. The increase in the number of enzyme and receptor 3D structures determined and the ability of combinatorial library screens to rapidly determine the binding affinities of candidate ligands to these targets.
3. Greatly increased sophistication of molecular modeling and QSAR analysis packages.
4. Progress in screening and matching algorithms, in particular, the ability to search all possible conformations of a compound in a database "on the fly" in order to determine if any match the query pharmacophore (Brint et al. 1987; Martin et al. 1988).

The ability to "mine" chemical structure databases is revolutionizing the search for new drugs. The cycle of search, synthesis, screen, and analysis of potentially thousands of new compounds can be reduced from years to months. It is confidently expected that most, if not all, new drugs coming to market in the next few years will have been created by a combination of some or all of these techniques.

Structure-Based Drug Design—Use of X-ray and NMR Structures

X-ray crystallography is a critical tool for any structure-based drug design program. This method provides the highest resolution information on the structures of both ligands and receptors, and ideally of the two bound together. Crystallography provides a static, time-averaged view of the molecule, providing good measures of bond lengths and angles as well as the included and excluded volumes in an active site or region when a ligand is bound or docked into that site. The disadvantages of classical crystallography include the fact that the solved structure may not represent the minimum energy conformation (particularly if it is a small molecule), that the observed structure typically represents only one of many possible conformations, and that high-quality crystals are needed to generate the data in the first place.

In an X-ray crystallography experiment, crystals of the molecule of interest are induced by a variety of methods, until high-quality, high-purity crystals of 0.1–0.2 mm lengths are produced. The crystals are placed in a diffractometer or synchrotron X-ray source and bombarded with X rays. In the case of proteins, the crystals (which contain on average 50% solvent and are therefore very fragile) are placed in a glass capillary in their "mother liquor" (solvent of crystallization) and flash frozen in order to preserve their structure during exposure to the X-ray beam. Typically the diffracted X rays are measured on an area detector connected to a computer with automatic data processing and data reduction software; this allows data sets to be obtained from the native crystal as well as from "heavy atom" derivatives (crystals with heavy atoms soaked in and hopefully bound to the protein of interest, that can then provide the necessary diffraction contrast in order to determine the native protein structure). In practice, an iterative procedure of data collection, data reduction, and structure refinement occurs, with a subset of the data used to generate an initial (low-resolution) electron density map, which is then refined using more data and other structure and sequence information in order to get the final structure at a resolution acceptable for drug design studies. (Typically most proteins are published initially at about 1.5–3Å resolution, which may or may not be sufficient for rational drug design.) It is important to remember that the X-ray crystal structure of a ligand-receptor complex is the time-averaged structure of a continuously moving system, and as such gives little indication of the flexibility or "strain" possible in the complex. Technological advances in crystallography include robotics-based crystallization procedures, improved area detectors for data collection, powerful and tuneable sources of X rays from synchrotrons, and more sophisticated software for data collection, reduction, analysis, and graphical display. There are currently over 2500 X-ray structures deposited in the Brookhaven Data Bank, April 1995 release (Bernstein et al. 1977).

A related technique that can provide complementary and sometimes competing information to X-ray crystallography is nuclear magnetic resonance (NMR). Both structure assignments (of peptides and small proteins) and solution dynamics can be explored. NMR probes the chemical environment around nuclei that exhibit magnetic spin (e.g., hydrogen atoms) by placing them in a strong magentic field and exciting them with a radio frequency pulse. The emitted frequency of the radio waves from the nuclei as they relax is shifted depending on their molecular environment, and these chemical shifts can be correlated with their 3D structure. In practice, the chemical shifts from protons (hydrogen ions) in a protein can be resolved only by using two- or three-dimensional NMR. A correlation spectroscopy (COSY) experiment gives peaks between protons that are covalently bonded via one or two atoms, while an NOE experiment (nuclear Overhauser effect) provides shifts that are between pairs of protons close together in space, even though they may not be from the same amino acid. By combining these two types of spectra with the amino acid sequence of the protein, and incorporating distance constraints, a single structure (or closely similar family of structures) may be derived, at least for smaller proteins (150 or fewer amino acids).

Even when the structure of the receptor (target) molecule has been elucidated at atomic resolution, a number of issues remain in selecting or designing an appropriate inhibitor (Kuntz 1992; Smellie et al. 1991; Sudarsanam et al. 1992). When a ligand is not present in the solved structure, one must be "docked" into the target site by considering the energy of ligand binding. Any calculation of the binding energy of the simulated docking of a ligand to a target must take into account the "cavitation" effect (the loss of bound solvent from the ligand and receptor sites), van der Waals/hydrophobic effects, electrostatic and induced electrostatic effects, conformational effects ("induced fit") in the target, and the entropic effect resulting from the restriction of several degrees of freedom such as the vibrational, rotational, and translational degrees of freedom in the ligand molecule. Since there is no method to accurately calculate each of these effects, approximations must be made, and in particular, entropic effects (evaluation over all states) are usually ignored. It should also be noted that multiple binding modes of the ligand to its target are also possible, as has been identified in an elastase-product complex. Free energy perturbation methods aim to remedy this important deficiency by simulating the physical process of determining the transfer of a drug from solution to its receptor binding site, as compared to the transfer of an analog. The process is simulated by "mutating" the drug into its analog, both at the binding site and in solution. In effect the initial drug molecule is run through a molecular dynamics simulation including several hundred solvent molecules, and the parameters of the simulation systematically altered from an initial 100% drug character to a final 100% analog character (van Gunsteren and Berendsen 1987; McCammon 1987). Free energy calculations for the competitive binding of benzamidine rather than fluorobenzamide to trypsin and of different thermolysin inhibitors to this enzyme have been shown to be within a single kilocalorie of the experimental value. These simulations underscore the importance of desolvation effects in the binding of ligand to target, and although as yet only single atom replacements have been attempted, complete side-chain replacements are hoped for in the future. (The latter are of considerably more interest to synthetic organic chemists designing new drugs.) The great advantage of structure-based design is that rather than trying to find molecules that adopt suitable geometries for a single binding mode, the entire receptor site can be explored, allowing a diversity of receptor–ligand interactions to be considered. A number of different methodologies have been adopted for docking ligand databases into known target structures (Sudarsaram et al. 1992; Lawrence and Davis 1992;

Desjarlais et al. 1988). One of these, the DOCK algorithm, is discussed in more detail below.

De novo Drug Design Based on Target 3D Structures

The most ambitious route to drug development is to create completely new compounds as drugs. These molecules may be based on existing inhibitors or antagonists, or created from scratch, atom by atom. At least three distinct approaches to *de novo* design (the design of novel compounds against a target based on structural and pharmacological information about that target) have emerged: directed design, random design, and grid-based design. For each of these, the *de novo* paradigm can be split into two phases: structure generation (either atom by atom or by linking together existing fragments and templates), and structure evaluation (whereby the structures are assessed and prioritized using a scoring scheme). Directed design has the user direct the process of creating novel molecules from a fragment library of appropriately selected fragments; random design uses a genetic (or other) algorithm to randomly mutate fragments and their combinations while simultaneously applying a selection criterion (binding energy) to evaluate the "fitness" of candidate ligands, while the grid-based approach calculates the interaction energy between 3D receptor structures and small fragment probes. Design in the latter case is accomplished by linking together favorable probe sites. Each of these methodologies has its own strengths and weaknesses, depending on whether one wishes to build molecules from linked fragments that have been matched to receptor site points, from linked fragments grown from a seed point using a potential energy function, or from linked fragments built using irregular lattices of previously docked molecules. *De novo* design is also possible through automatic lead modification or linking of random fragments coupled with constraints derived from QSAR analysis. One of the most exciting developments in *de novo* design is the use of the known structure of a receptor as a "virtual" screen against a combinatorial library, allowing the *de novo* design methodology to generate diversity focussed toward a specific target. Although *de novo* design has not lived up to its early promise entirely, there are a number of molecules in clinical trials that have been assisted by *de novo* design philosophies. Agouron, Inc. has three compounds in clinical trials of inhibitors to thymidilate kinase (TK), an important target in anticancer drug design, and one compound that is an inhibitor to HIV protease, all designed using the principles described above. Other examples of *de novo* design include the HIV protease and reverse transcriptase inhibitors from Abbott and Merck and PNP inhibitors from BioCryst (see the sidebar).

APPLICATIONS OF STRUCTURE-BASED DRUG DESIGN

Rational Design of Potent Antagonists to the Human Growth Hormone (hGH) Receptor (Genentech, Inc., San Francisco, CA)

High-resolution mutational analysis coupled with X-ray crystallography have defined two sites on human growth hormone (hGH) that bind to two molecules of the extracellular domain of its receptor (hGHbp). Dimerization of the receptor occurs sequentially, with one receptor binding to site 1 on hGH and the second consequently on site 2 of hGH and at a site on the first hGHbp molecule. A cell-based signaling assay for hGH was constructed by creating a hybrid receptor containing the extracellular domain of hGHpb linked to the transmembrane and intracellular domains of the murine granulocyte colony-stimulating factor receptor. This molecule was expressed in a myeloid leukemia,cell line (FDC-P1) and shown to be functional by addition of hGH, which causes proliferation of these cells. X-ray crystallography of the hGH-receptor complex shows that residues Lysine-172 (K172) and Phenylalanine-176 (F176) are located in site 1 of hGH while Glycine-120 (G120) is located in site 2 and makes a van der Waals contact with residue Tryptophan-104 (W104) of the receptor. A mutant G120R (Glycine replaced by Arginine at position 120) retains a functional site 1 but a sterically blocked site 2. This variant does not affect cell proliferation in the cell line suggesting that binding to either site 1 or 2 is necessary but not sufficient for promoting cell proliferation. Mutants blocking site 2 were hypothesized to be antagonists to cell proliferation—as expected, site 2 mutants antagonized hGH, while site 1 mutants (either at positions K172 or F176) did not. hGH agonism and antagonism to its receptor is illustrated in Fig. 4.6.

The design of potent antagonists to the hGH receptor may be useful in the clinical treatment of hGH excess acromegaly (excessive growth). A transgenic mouse that expresses large amounts of bovine hGH altered in site 2 has been shown to produce dwarf mice. The discovery that hGH binds to two receptors in a sequential fashion was not expected and helps to explain a number of pharmacokinetic features of these molecules. The mechanism-based strategy for the design of potent antagonists to hGH may be applied to other hormones such as prolactin, placental lactogen, the interleukins, colony-stimulating factors, the hematopoietines and the cytokines, assuming that they too require sequential binding of two receptors to a single hormone for signaling to occur.

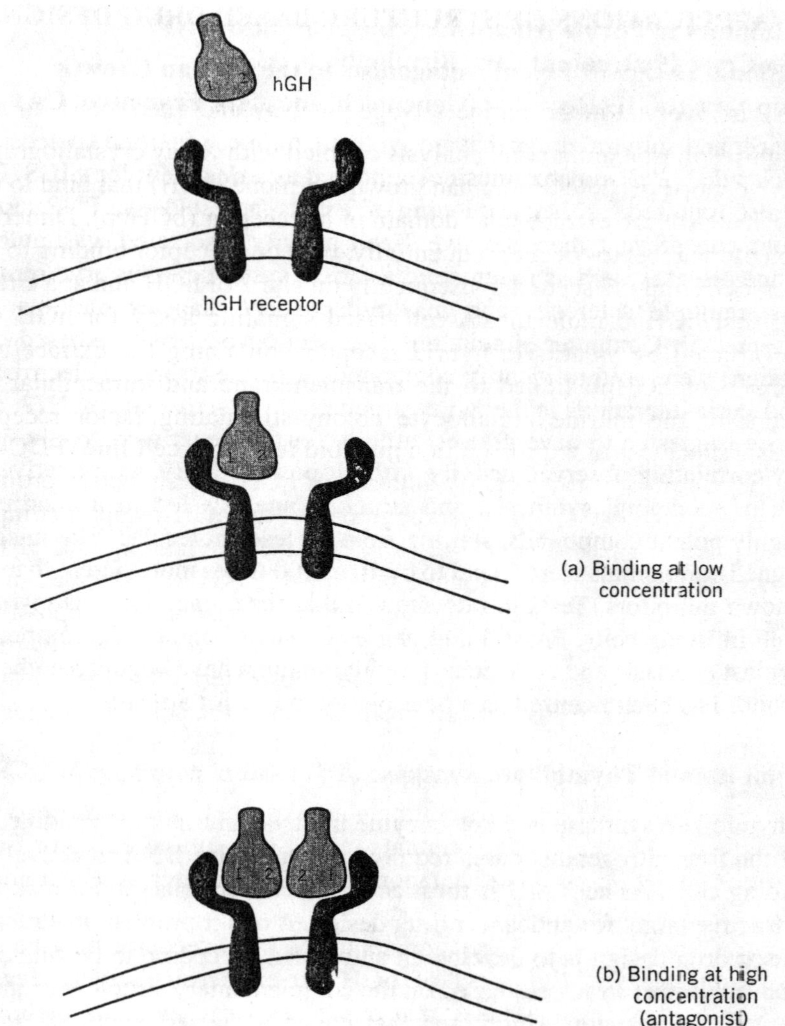

Figure 4.6. Human growth hormone (hGH)—receptor structure and mode of action. The hGH is postulated to act as both an agonist (at low concentrations) and antagonist (at high concentrations) at its receptor. In the sequential dimerization model, at low concentrations, hGH binds first at site 1 and subsequently at site 2 (as indicated) to produce an active hGH-$(hGHR)_2$ complex. At high concentrations, hGH saturates the receptor through site 1 interactions and acts as an antagonist.

Inhibitors to Purine Nucleoside Phosphorylase (PNP) (BioCryst Pharmaceuticals, Birmingham, AL)

PNP is involved in the purine salvage pathway and interferes with anticancer and antiviral drugs that are synthetic mimics of purine nucleosides, such a ddI (2′,3′-dideoxyinosine) approved as a treatment for AIDS. PNP is also required for the functioning of T cells, and selective PNP supressions can prevent the excessive T-cell activity associated with autoimmune diseases such as rheumatoid arthritis, systemic lupus erythromatosus, multiple sclerosis, and insulin-dependent diabetes mellitus (see Chapter 5). Computer simulations (see section on structure-based drug design) were used to identify compounds with the strongest electrostatic and steric interations to the X-ray structure of the active site of PNP; only those suggested to have greatest affinity for the target were synthesized. By correlating observed activity with proposed affinity, an interative cycle of modeling, synthesis, and structure analysis led to a handful of highly potent compounds, starting from 60 lead molecules. The final designed compounds were found to be 10 to 100 times more potent than any known inhibitors. Tests in rats showed that the compounds also worked well in living cells. Phase I and phase II trials of one of the compounds against psoriasis and cutaneous T-cell lymphoma have begun. Another inhibitor has been licensed as a possible treatment for arthritis.

Inhibitors of Thymidilate Synthase (TS) (Kuntz et al., 1992; UCSF)

Thymidilate synthase is a key enzyme in the synthesis of thymidine, one of the four nitrogenous bases required to synthesize DNA in actively dividing cells. As such, TS is most active in cancer cells, and it makes an attractive target for anticancer drug design. A major problem in structure-based drug design is to develop an automated procedure to postulate ligand (inhibitor) structures based on the complementary structure of the receptor (active) site. Algorithms that create a "negative image" of the receptor site, fit putative inhibitors into that site, and measure the quality of the fit have now been developed. The UCSF DOCK program fills the receptor site with overlapping spheres, using the sphere centers to generate the negative image. Rapid procedures compare the overlapping geometries of putative inhibitors, with scoring based on a steric complementarity or a full intermolecular force field. High scoring matches from a database can be used as template structures for the design of novel molecules. Experience with the DOCK program shows that 2–20% of the compounds discovered show inhibition in the micromolar range. DOCK was used to find inhibitors to thymidylate synthase (TS) by searching the available Chemicals Database (ACD). DOCK not only retrieved the substrate and several known inhibitors but also proposed putative inhibitors previously not known to bind the enzyme. Three of these compounds inhibited the enzyme at submillimolar concentrations. X-ray structure analysis of one of these inhibitors complexed to the enzyme revealed a novel conformation of the substrate due to the presence of a counterion, suggesting another binding region in the active site that could be used as a search template. This region was again probed with DOCK with the resultant identification of a series of phenolphthalien analogs that also inhibit TS in the 1–30 micromolar range. These novel inhibitors do *not* resemble the substrates of the enzyme. X-ray structure analysis of one of these new inhibitors complexed with the enzyme confirms that its binding in the target site is as predicted by DOCK.

The use of DOCK and other programs in optimizing leads requires obtaining the proper ligand conformation as well as discriminating among several proposed modes of interaction. Assumptions in searching databases include assuming that either the receptor or the ligand is rigid, neglecting bound water molecules and counterions, and simplifying evaluations of interaction energies. The effect of desolvation of the ligand and receptor surfaces on binding and delineation of accurate binding geometries should allow for the calculation of accurate free energies. The speedup in screening means that screens that took one to two years to complete can now be simulated by computer in days to weeks.

Structure-Based Drug Design of HIV Protease Inhibitors (Merck, Upjohn, Abbott; Roche, Agouron, Vertex)

A number of academic and commercial establishments have been attempting to treat HIV-1 infection, the primary causal agent of AIDS, by developing inhibitors to the virally encoded enzymes needed for viral replication and assembly. Because HIV-1 produces the gag and gag-pol gene products as inactive polyproteins, a viral protease is needed to cleave these molecules into the viral proteins p17, p24, p7, p6, and the essential viral enzymes such as reverse transcriptase and integrase. Inhibition of HIV-1 protease results in the budding and fusion of immature noninfectious particles, interrupting completion of the viral life cycle. Unlike reverse transcriptase inhibitors, which are only effective in blocking viral replication in acutely infected cells, antiprotease drugs inhibit production of infectious virus in both acutely and chronically infected cells.

The HIV-1 protease molecule has had its X-ray structure determined in the absence and the presence of a number of transition-state based inhibitors (Wlodawer et al. 1989; Erickson et al. 1990). It is a homodimer consisting of two identical 99 amino-acid long peptides with a "signature" Asp-Thr-Gly sequence that includes the catalytic Asp residue at the interface of the two domains (positions 25–27 and 125–127 in the two peptide chains—the second domain being numbered sequentially after the first). Development of potent nonpeptide inhibitors to this enzyme has required efficient computer-based models that measure the affinity of candidate compounds to the target (Hoog et al. 1995). 3D database searching has also been utilized to find synthetic frameworks that can be used as the starting point in the design of nonpeptide inhibitors (Wlodawer 1994). Predictions of affinity have included simulations (e.g. Monte Carlo, molecular dynamics), as well as correlation paradigms (QSARs and learning algorithms). These methods are considered essential for successful *de novo* design of nonpeptide inhibitors in order to optimize and score affinities, establish priorities for synthesis, and compare against experimental data in order to refine the model building. Marshall and coworkers (Washington University) have used both in-house algorithms and commercial software to generate pharmacophores for HIV-1 protease and combined them with 3D database searches and *de novo* "splicing" algorithms to generate novel compounds based on cyclic ring template structures. Substituents recovered from the database were used to create numerous derivatives that were in turn used as the input for SAR data sets. 3D-QSAR using CoMFA-based and learning-based algorithms (coded neural nets) validated the predictive power of this technique in the presence of the receptor-site 3D structure, suggesting that using the receptor to generate an alignment rule for 3D-QSAR is a valuable adjunct in predicting the affinity of related molecules (Green and Marshall 1995).

A large group at Merck (Lam et al. 1994) has created a number of HIV-1 protease inhibitors based on the structure of nonpeptide molecules designed around cyclic urea scaffolds. 3D-QSAR analysis of these compounds by back-propagation neural networks and by CoMFA suggest that these related methods are powerful screening functions for evaluating the substituent properties of these trial compounds. Important functional group properties seem to include the electrostatic interaction energy, a steric interaction energy, steric fit, partition coefficient, lipophilic contact surface area, hydrophilic contact surface area, and ligand conformational enthalpy (the "strain" on the ligand imposed by binding). In a related approach Wang et al. (1995) have used the technique of solid-phase combinatorial chemistry described in the previous chapter to generate c2-symmetric inhibitors constrained by the known X-ray structure. Over 300 compounds, based on diamino and diamino alcohol cores, were generated in order to identify submicromolar inhibitors.

The principal issue with the development of these inhibitors seems to be the capability of the protease to tolerate single-residue (point) mutations that render the compounds ineffective yet do not significantly affect the activity of the enzyme (Moore and Huffman 1993). Classification of the type and position of these mutations is essential in determining the "mix" of inhibitors needed if the virus is to be thwarted. Trial drugs from Merck (MK-639, commercial name Indinavir) and from Abbott (ABT-538, commercial name Ritonavir) show initial antiviral activity as observed by studying the population dynamics of HIV and of CD4 cells in circulation, leading to rapid reduction of viral protease replication and restoration of CD4 cell counts. However, within 6 to 12 months of therapy with these inhibitors, antiviral activity is essentially dissipated in conjunction with the acquisition of a highly resistant strain of the virus. The mutability of HIV protease is remarkable. At least 20 amino acids of the 99 in each dimer can undergo mutation, while maintaining enzyme activity when faced with selective pressure from different inhibitors (Mellors et al. 1995). These mutations are centered either around the active site pocket delineated by X-ray crystallography or in regions of the protease that can undergo conformational changes and indirectly affect binding. Still unanswered is the question of whether multiple inhibitor therapy can be used to prevent antiviral activity over long time periods. Nonetheless, a number of HIV protease inhibitors have entered clinical trials, and early results suggest significant therapeutic benefit, particularly when used in combination with other drugs. Indeed the first rationally designed inhibitor to HIV-1 and HIV-2 protease (Saquinivir from Hoffman LaRoche) has recently been approved for clinical use in the United States, and several others are on the way to approval (Vella 1994).

4.5. SUMMARY

In the past, drugs were created with almost complete naivete regarding the molecular mechanisms of the underlying molecular machinery involved in their potency. Recent developments have underlied the critical importance of three dimensionality in molecular recognition and discrimination. Even in the absence of the 3D structure of the target molecule, drug design that takes into account the 3D flexibility of candidate ligands has helped revolutionize the discovery of new lead compounds. Integrated approaches that combine modeling with experimental methodologies such a QSAR and other related techniques may be especially powerful in forecasting the affinity of novel molecules designed this way.

Structure-based drug design is a rapidly evolving methodology that has found a niche in every large pharmaceutical company as well as at many smaller biotechnology companies. Indeed several companies (Agouron, Vertex, BioCryst) have been formed with the express intention of using solely this technology to develop new drugs. Although some of the early expectations of this methodology have been somewhat exaggerated, there is no doubt overall that the use of computers in drug discovery, design, and optimization can be an extraordinarily powerful tool. The developments in combinatorial chemistry and molecular diversity described in the previous chapter (sometimes termed "irrational" drug design) are unlikely to reduce the importance of this technology. On the contrary, the best use of the two methods (one experimental, the other computational) may be in combination, enabling drug designers to simulate, synthesize and screen large numbers of compounds in a rational manner. The expectation is that potent new drugs will for the first time be developed rapidly *and* effectively (see Peisach et al. 1995, for an example of this approach).

REFERENCES

Allen, M. S., Skolnick, P., and Cook, J. M. 1992. Synthesis of novel 2-phenyl-2H-pyrazolo[4,3-c]isoquinolin-3-ols: Topological comparisons with analogues of 2-phenyl-2,5-dihydropyrazolo[4,3-c]quinolin-3(3H)-ones at benzodiazepine receptors. *Journal of Medicinal Chemistry* 35(2):368–372.

Appelt, K., Bacquet, R. J., Bartlett, C. A., Booth, C. L. J., Freer, S. T., Fuhry, M. A. M., Gehring, M. R., Herrmann, S. M., and Howl, 1991. Design of enzyme inhibitors using iterative protein crystallographic analysis. *Journal of Medicinal Chemistry* 34(7):1925–1934.

Bernstein, F. C., Koetzle, T. F., Williams, G. J., Meyer, E. E., Jr., Brice, M. D., Rodgers, J. R., Kennard, O., Shimanouchi, T., and Tasumi, M. 1977. The Protein Data Bank: A computer-based archival file for macromolecular structures. *Journal of Molecular Biology* 112(3):535–542.

Blow, D. M., Fersht A. R., and Winter G. (eds.). 1986. *Design, Construction and Properties of Novel Protein Molecules.* London: Royal Society.

Blundell, T. L., Elliott, G., Gardner, S. P., Hubbard, T., Islam, S., Johnson, M., Mantafounis, D., and Murray-Rust, P. 1989. Protein engineering and design. *Philosophical Transactions of the Royal Society of London B Biological Sciences* 324(1224):447–460

Blundell, T. L., 1994. Problems and solutions in protein engineering-towards rational design. *Trends in Biotechnology* 12(5):145–148.

Brint, A. T., and Willett, P. J. 1987. Upperbound procedures for the identification of similar three-dimensional chemical structures. *Computer-Aided Molecular Design* 2(4):311–320.

Burgen, A. S. V., Roberts, G. C. K., and Tute, M. S. (eds.). *Topics in Molecular Pharmacology,* vol. 3: *Molecular Graphics and Drug Design.* 365pp. Amsterdam: Elsevier Science Publishers.

Chen, S., Chrusciel, R. A., Nakanishi, H., Raktabutr, A., Johnson, M. E., Sato, A., Weiner, D., Hoxie, J., Saragovi, H. U., Greene, M. I., et al. 1992. Design and synthesis of a CD4 beta-turn mimetic that inhibits human immunodeficiency virus envelope glycoprotein gp120 binding and infection of human lymphoctyes. *Proceedings of the National Academy of Sciences* 89(13):5872–5876.

Chew, C., Villar, H. O., and Loew, G. H. 1991. Theoretical study of the flexibility and solution conformation of the cyclic opioid peptides [D-Pen2,D-Pen5]enkephalin and [D-Pen2,L-Pen5]enkephalin. *Molecular Pharmacology* 39(4):502–510.

Cramer, R. D., III, Patterson, D. E., and Bunce, J. D. 1988. Comparative molecular field analysis (CoMFA): 1. Effect of shape on binding of steroids to carrier proteins. *Journal of the American Chemical Society.* 110(18):5959–5967.

Cramer, R. D., III, and Wold S. B. 1991. U.S. Patent No. 5,025,388. U.S. Patent Office, Alexandria, VA.

Dammkoehler, R. A., Karasek, S. F., Shands, E. F. B., and Marshall, G. R. 1989. Constrained search of conformational hyperspace. *Journal of Computer-Aided Molecular Design* 3(1):3–22.

de Vos, A. M., Ultsch, M., and Kossiakoff, A. A. 1992. Human growth hormone and extracellular domain of its receptor: Crystal structure of the complex. *Science* 255(5042):306–312.

Dean, P. M. 1987. *Molecular Foundations of Drug-Receptor Interactions.* Cambridge: Cambridge University Press.

Debnath, A. K., Hansch, C., Kim, K. H., and Martin, Y. C. 1993. Mechanistic interpretation of the genotoxicity of nitrofurans (antibacterial agents) using quantitative structure-activity relationships and comparative molecular field analysis. *Journal of Medicinal Chemistry* 36(8):1007–1016.

DePriest, S. A., Mayer, D., Naylor, C. B., and Marshall, G. R. 1993. 3D-QSAR of angiotensin-converting enzyme and thermolysin inhibitors: A comparison of CoMFA models based on deduced and experimentally determined active site geometries. *Journal of the American Chemical Society* 115(13):5372–5384.

Desjarlais, R. L., Sheridan, R. P., Seibel, G. L., Dixon, J. S., Kuntz, I. D., and Venkataraghavan, R. 1988. Using shape complementarity as an initial screen in designing ligands for a receptor binding site of known three-dimensional structure. *Journal of Medicinal Chemistry* 31(4):722–729.

Dougherty, D. A., and Stauffer, D. A. 1990. Acetylcholine binding by a synthetic receptor: Implications for biological recognition. *Science* 250(4987):1558–1560.

Erickson, J., Neidhart, D. J., VanDrie, J., Kempf, D. J., Wang, X. C., Norbeck, D. W., Plattner, J. J., Rittenhouse, J. W., Turon, M., Wideburg, N., et al. 1990. Design, activity, and 2.8 A crystal structure of a C2 symmetric inhibitor complexed to HIV-1 protease. *Science* 249(4968):527–533. Abstract available.

Evans, B. E., Leighton, J. L., Rittle, K. E., Gilbert, K. F., Lundell, G. F., Gould, N. P., Hobbs, D. W., Dipardo, R. M., Veber, D. F., et al. 1986. Orally active, nonpeptide oxytocin antagonists. *Journal of Medicinal Chemistry* 35(21): 3919–3927.

Findlay, J., and Eliopoulos, E. 1990. Three-dimensional modelling of G-protein-linked receptors. *Trends in Pharmacological Sciences* 11(12):492–499.

Gay, D., Maddon, P., Sekaly, R., Talle, M. A., Godfrey, M., Long, E., Goldstein, G., Chess, L., Axel, R., Kappler, J., et al. 1987. Functional interaction between human T-cell protein CD4 and the major histocompatibility complex HLA-DR antigen. *Nature* 328(6131):626–629.

Ghose, A. K., Crippen, G. M., Revankar, G. R., Mckernan, P. A., Smee, D. F., Robins, 1989. Analysis of the in vitro antiviral activity in certain ribonucleosides against parainfluenza virus using a novel computer aided receptor modeling procedure. *Journal of Medicinal Chemistry* 32(4):746–756,

Greco, G., Novellino, E., Silipo, C., and Vittoria, A. 1992. Study of benzodiazepines receptor sites using a combined QSAR-CoMFA approach. *Quantitative Structure-Activity Relationships* 11(4):461–477.

Green, S. M., and Marshall, G. R. 1995. 3D-QSAR: A current perspective. *Trends in Pharmacological Sciences* 16(9):285–291.

Goetz, M. A., Monaghan, R. L., Chang, R. S. L., Ondeyka, J., Chen, T. B., and Lotti, V. J. 1986. Novel cholecystokinin antagonists from Aspergillus alliaceus: I. Fermentation, isolation, and biological properties. *Journal of Antibiotics* (Tokyo) 1986 41(7):875–877.

Goodford, P. J. 1985. A computational procedure for determining energetically favorable binding sites on biologically important macromolecules. *Journal of Medicinal Chemistry* 28(7):849–857.

Gorin, F. A., Balasubramanian, T. M., Barry, C. D., and Marshall, G. R. 1978. Elucidation of the receptor-bound conformation of the enkephalins. *Journal of Supramolecular Structure* 9(1):27–39.

Gros, P., Van Gunsteren, W. F., and Hol, W. G. J. 1990. Inclusion of thermal motion in crystallographic structures by restrained molecular dynamics. *Science* 249(4973):1149–1152.

Gund P. 1974. *In Progress in Molecular and Subcellular Biology,* vol. 5, F. E. Hahn (ed.) New York: Springer Verlag, pp. 117–123.

Hancsh, C., and Leo, A. 1979. *Substituent Constants for Correlation Analysis in Chemistry and Biology.* New York: Wiley-Interscience.

Hoog, S. S., Zho, B., Winborne, E., Fisher, S., Green, D. W., Desjarlais, R. L., Newlander, K. A., and Callahan, A. A 1995. A check on a rational drug design: Crystal structure of complex of human immunodeficiency virus type 1 protease with a novel gamma-turn mimetic. *Journal of Medicinal Chemistry* 38(17): 3246–3252.

Hughes, J., Smith, T. W., Kosterlitz, H. W., Fothergill, L. A., Morgan, B. A., and Morris, H. R. 1975. Identification of two related pentapeptides from the brain with potent opiate agonist activity. *Nature* 258(5536):577–580.

Kim, K. H., and Martin, Y. C. 1991. Direct prediction of linear free energy substituent effects from 3D structures using comparative molecular field analysis: 1. Electronic effects of substituted benzoic acids. *Journal of Organic Chemistry* 56(8):2723–2729.

Kuntz, I. D., Meng, E. C., Shoichet, B. K., Bodian, D. L., and Roe, D. C. 1992. Advances in computer-aided design of enzyme inhibitors. *Abstracts of Papers American Chemical Society* 203(1–3):MED16.

Kuntz, I. D. 1992. Structure-based strategies for drug design and discovery. *Science* 257(5073):1078–1082.

Lawrence, M. C., and Davis, P. C. 1992. CLIX: A search algorithm for finding novel ligands capable of binding proteins of known three-dimensional structure. *Proteins Structure Function and Genetics* 12(1):31–41.

Lam, P. Y., Jadhav, P. K., Eyermann, C. J., Hodge, C. N., Ru, Y., Bacheler, L. T., Meek, J. L., Otto, M. J., Rayner, M. M., Wong, Y. N., et al. 1994. Rational design of potent, bioavailable, nonpeptide cyclic ureas as HIV protease inhibitors. *Science* 263(5145):380–384.

Leach, A. R., Prout, K., and Dolata, D. P. 1990. Automated conformation analysis: Algorithms for the efficient construction of low-energy conformations. *Journal of Computer-Aided Molecular Design* 4(3):271–282.

Lewis, R. A., Roe, D. C., Huang, C., Ferrin, T. E., Langridge, R., and Kuntz, I. D. 1992. Automated site-directed drug design using molecular lattices. *Journal of Molecular Graphics* 10(2):66–78, 106.

Loew, G. H., Lawson, J. A., Toll, L., Polgar, W., and Uyeno, E. T. 1991. Structure-activity studies of morphine fragments: II. Synthesis, opiate receptor binding, analgetic activity and conformational studies of 2-R-2-(hydroxybenzyl)piperidines. *European Journal of Medicinal Chemistry* 26(8):763–774.

Magnusson, S. L., et al. 1976. In *Miami Midwinter Symposia. Vol. 2:* Proteolysis and Physiological Regulation, D. W. Ribbons and K. Brew (eds.), New York: Academic Press, pp. 203–239.

Marshall, G. R. et al. 1979. In *Computer-Assisted Drug Design,* eds. E. C. Olson and R. E. Christofferson, (eds.), *ACS Symposium Series* 112. New York: American Chemical Society, p 205.

Marshall, G. R., and Cramer, R. D., III. 1988. Three-dimensional structure-activity relationships. *Trends in Pharmacological Sciences* 9(8):285–289.

Marshall, G. R., and Motoc, I. 1986. Approaches to the conformation of the drug bound to the receptor. G. C. K, Roberts, and M. S. Tute (eds.). A. S. V. Burgen, In *Topics in Molecular Pharmacology: Molecular Graphics and Drug Design,* vol. 3. Amsterdam: Elsevier Science Publishers, pp. 115–156.

Martin, Y. C., Danaher, E. B., May, C. S., and Weininger, D. 1988. MENTHOR, a database system for the storage and retrieval of three-dimensional molecular structures and associated data searchable by substructural, biologic, physical, or geometric properties. *Journal of Computer-Aided Molecular Design* 2(1):15–30.

McCammon, J. A. 1987. Computer-aided molecular design. *Science* 238 (4826):486–491.

Mellors, J. W., Bazmi, H. Z., Schinazi, R. F., Roy, B. M., Hsiou, Y., Arnold, E., Weir, J., and Mayers. 1995. Novel mutations in retroviral protease of human immunodeficiency virus type 1 reduce susceptibility to saquinivir in laboratory and clinical isolates. *Antimicrobial Agents and Chemotherapy* 39(5):1087–1092.

Moore, M. L., and Huffman W. L. 1993. Substrate-based inhibitors of HIV-1 protease. *Perspectives in Drug Discovery and Design* 1(1):85–108.

Motoc, I., Dammkoehler, R. A., and Marshall, G. R. 1986. In *Mathematical and Computational Concepts in Chemistry.* N Trinajstic, (ed.). Chichester: Horwood Ltd., p. 222.

Nakanishi, H., Chrusciel, R. A., Shen, R., Bertenshaw, S., Johnson, M. E., Rydel, T. J., Tulinsky, A., and Kahn, M. 1992. Peptide mimetics of the thrombin-bound structure of fibrinopeptide A. *Proceedings of the National Academy of Sciences* 89(5):1705–1709.

Peisach, E., Casebier, D., Gallion, S. L., Furth, P., Petsko, G. A., Hogan, J. C., Jr., Ringe, D. 1995 Interaction of a peptidomimetic aminimide inhibitor with elastase. *Science* 269(5220):66–69.

Ripka, W., Sipio, W. J., and Galbraith, W. G. Molecular modeling in the design of phospholipase a2 inhibitors. *Journal of Cellular Biochemistry,* suppl. 12E:83.

Rees, A. R., Staunton, D., Webster, D. M., Searle, S. J., Henry, A. H., and Pedersen, J. T. 1994. Antibody design: Beyond the natural limits. *Trends in Biotechnology* 12(5):199–206.

Rose, G. D. 1988. Protein folding: New twists. *Bio-Technology* 6(2):167–171.

Rusinko J. H., and Pearlman D. L. 1988. *CONCORD.* St. Louis, MO: Tripos Associates Inc.

Saragovi, H., Fitzpatrick, D., Raktabutr, A., Nakanishi, H., Kahn, M., and Greene, M. I. 1991. Design and synthesis of a mimetic from an antibody complementarity-determining region. *Science* 253(5021):792–795.

Sattentau, Q. J., and Weiss, R. A. 1988. The CD4 antigen: Physiological ligand and HIV receptor. *Cell,* 52(5):631–633.

Sleckman, B. P., Peterson, A., Jones W. K., Foran, J. A., Greenstein, J. L., Seed, B., Burakoff, S. J. 1987. Expression and function of CD4 in a murine T-cell hybridoma. *Nature* 328(6128):351–353.

Smellie, A. S., Crippen, G. M., Richards, W. G. 1991. Fast drug-receptor mapping by site-directed distances: A novel method of predicting new pharmacological leads. *Journal of Chemical Information and Computer Sciences* 31 (3):386–392.

Sudarsanam, S., Virca, G. D., March, C. J., and Srinivasan, S. 1992. An approach to computer-aided inhibitor design: Application to cathepsin L. *Journal of Computer-Aided Molecular Design* 6(3):223–233.

Van Gunsteren, W. F., and Berendsen, H. C. 1987. Conformation search by potential energy annealing: Algorithm and application to cyclosporin A. *Journal of Computer-Aided Molecular Design* 6(2):97–112.

Wang, G., Li, S., Wideburg, N., Krafft, G. A., and Kempf, D. J. 1995. Synthetic chemical diversity: Solid phase synthesis of libraries of C-2 symmetric inhibitors of HIV protease containing diamino diol and diamino alcohol cores. *Journal of Medicinal Chemistry* 38(16):2995–3002.

Wlodawer, A., Miller, M., Jaskolski, M., Sathyanarayana, B. K., Baldwin, E., Weber, I. T., 1989. Conserved folding in retroviral proteases crystal structure of a synthetic hiv-1 protease. *Science* 245(4918):616–621.

Wlodawer, A. 1994. Rational drug design: The proteinase inhibitors. *Pharmacotherapy,* 14(6, pt. 2):9S–20S.

Wu, T. P., Yee, V., Tulinsky, A., Chrusciel, R. A., Nakanishi, H., Shen, R., Priebe, C., and Kahn, M. 1993. The structure of a designed peptidomimetic inhibitor complex of alpha-thrombin. *Protein Engineering* 6(5):471–478.

5

RE-ENGINEERING THE IMMUNE RESPONSE

5.1. AN OVERVIEW OF THE IMMUNE RESPONSE

The previous chapters have described the design and development of pharmaceutical drugs and therapeutic agents such as genes to correct genetic defects and to combat disease. There are some diseases, however, that are the result of the presence of many genes and are also influenced by the environment (the so-called multifactorial polygenic diseases). Examples of such diseases include insulin dependent diabetes mellitus (IDDM), rheumatoid arthritis (RA), and multiple sclerosis (MS). This chapter will describe therapeutic strategies that can be taken to combat complex diseases that result when the immune system goes awry.

The immune system is responsible for protecting the body from foreign pathogenic attack. Since most pathogens are built from the same protein-based structures as host cells, the immune system has developed novel methods that allow discrimination between self and foreign antigens. The acceptance or unresponsiveness to self antigens is referred to as *tolerance.* Failure of tolerance leads to an immunological assault on "self" tissues resulting in autoimmune diseases such as IDDM, MS, and many others (see Table 5.1).

An excessive response of the immune system can also result in immune pathologies, such as the inflammatory states seen with allergies. In recent years the molecular basis for specific antigen recognition by B and T cells has been elucidated, and the molecular mechanisms of T-cell activation defined. These studies have made it possible to design various immune inter-

Table 5.1. Common HLA-Linked autoimmune disorders

Disease
Rheumatoid arthritis
Insulin dependent diabetes mellitus
Multiple sclerosis
Pemphigus vulgaris
Chronic active hepatitis
Sjogren's syndrome
Celiac disease
Ankylosing spondylitis
Graves' disease
Hasimoto's thyroiditis
Systemic lupus erythematosus

vention strategies that re-engineer the immune system to prevent autoimmune phenomena and dampen the hyperactivity of the allergic response. Some of these strategies may also be viable to be used as therapies against transplantation rejection, since the mediators of rejection are the same as those controlling autoimmunity and allergy.

This chapter will highlight experimental therapeutic strategies used to prevent disease in two animal models, experimental allergic encephalomyelitis (EAE) (an MS-like state in mice), and diabetes in the nonobese diabetic (NOD) mouse. The lessons learned from these models relating to specificity, efficacy, and applicability of different strategies will be discussed. Finally, the future directions and prospects of eventually using immune intervention in human patients will be analyzed.

5.2. ORGANS AND CELLS OF THE IMMUNE SYSTEM*

The tissues of the immune system can be divided into two groups: the primary and secondary lymphoid organs. The primary organs are the site of generation of immunocompetent cells from precursors. Development of a single stem cell population in the bone-marrow produces all blood cells, including the B and T lymphocytes, red blood cells, and the other immune cells (see below). T-cell development occurs in the thymus. Secondary organs are the sites where mature lymphocytes respond to foreign antigens. These organs are distributed to serve different parts of the body. Lymph nodes are specialized structures situated along lymphatic channels around the body

*For a comprehensive review of basic immunology, the reader is referred to Abbas et al. (1995).

that provide the immunological milieu to survey for and react to foreign antigenic material in the lymphatic drainage from most intracellular spaces. Other lymph-node-like structures such as the mesenteric node, the appendix, and Peyer's patches drain the intestines. The spleen is the major site of immune response to blood-borne antigens.

The organs are traversed by a variety of immune cells that circulate in the blood and the lymph to reach almost all parts of the body. The different cells are listed below:

Lymphocytes	T cells
	B cells
Natural killer cells	NK cells
Phagocytes	Monocytes/macrophages
	Microglia (central nervous system)
	Kuppfer cells (liver)
	Alveolar macrophages (lung)
	Dendritic cells (Langerhans cells)
Granulocytes	Neutrophils
	Eosinophils
	Basophils (mast cells)

Lymphocytes form the arm of the immune response capable of specifically recognizing antigen. This is mediated by receptors on their cell surface. In the case of B cells, the receptors are immunoglobulins (Ig). Activation of B cells leads to their differentiation into antibody (Ab) secreting plasma cells. These antibodies, composed of two heavy chains and two light chains, are identical to the surface-linked immunoglobulins on the triggered B cell. Similarly for T cells, the receptor is the T-cell receptor (TCR). TCRs are disulphide-linked cell surface heterodimers. Ninety percent of the T cells bear alpha/beta heterodimers which predominate in the blood and secondary lymphoid organs, whereas gamma/delta T cells are present mostly in epithelial cells of the skin, gut, and genitourital tracts. Mechanisms exist to generate a vast array of Igs and TCRs (see the sidebar below), and each cell expresses only one Ig or TCR.

Alpha/Beta TCRs recognize small peptide fragments, derived from pathogens, presented to them by major histocompatibility complex (MHC) molecules on virally infected or so-called professional antigen-presenting cells. MHC class I molecules present antigen to T cells that are mainly killer cells and are thus called cytotoxic T lymphocytes (CTLs). These T cells carry a co-receptor that increases the affinity for MHC class I molecules and are thus often referred to as *CD8 cytotoxic T cells*. Similarly MHC class II molecules present antigens to a T-cell population that promotes the function of

The Arrangement of Antibody and T-Cell Receptor Genes (see Fig. 4.2 in Chapter 4)

A human being or animal can be exposed to numerous pathogens during its lifespan. These pathogens can be bacterial, viral, or parasitic, and they are composed of an almost limitless array of different proteins. Antibodies and T-cell receptors are the combating weapons the immune system utilizes to ward off these pathogens. To allow these arms of the immune response to detect as many variations of pathogens as possible, the immune system has evolved novel ways to generate large numbers of antibodies and T-cell receptors. Antibodies are composed of two identical heavy (H) and two identical light (L) chains held together by disulphide linkages. TCRs are composed of two chains designated alpha and beta. Antibodies are composed of a constant region responsible for effector function and a variable region that forms the antigen-binding site of the molecule. The antigen-binding regions are at the tips of the "Y" and are formed by heavy and light chain contributions. Similarly TCRs have a MHC-peptide-recognizing domain that is exposed at the cell surface away from the constant, membrane-traversing segments. In both cases separate gene fragments code for the distinct domains of each chain almost like exons. For example, separate variable and constant gene fragments exist for the chains of TCR and antibodies. These variable segments must be joined to other segments called D (diversity) regions and J (junctional) regions before they are joined to the constant segments. These events occur by a process known as *recombination*. Because numerous (hundreds for the variable segments) of each of these segments exist, their random combination can generate a large number of different "whole" genes. For example, in the mouse there are 200–300 VH genes, 10 D regions, and 4 JH segments. In addition, at the VDJ junction, further diversity is generated by the addition of random nucleotides. This, together with the hypermutation that occurs in antibody genes and the fact that both chains contribute to antigen binding, means that a total of 2.7×10^{28} different antibodies can be generated. A similarly large number of TCRs can be generated from a limited number of gene sections. D segments are only present in heavy chains and TCR beta chains. There are also different C regions, and the one used determines the isotype and therefore function of the antibody (IgA or IgG, etc.). Each B cell expresses only one antibody, and each T cell only one receptor. It is possible to isolate antibodies against an antigen that originate from one B-cell clone. Such antibodies are called *monoclonal antibodies*. Every antibody in such a population is identical with the exact same specificity. Thus monoclonal antibodies are very useful tools (refer to Section 4.2).

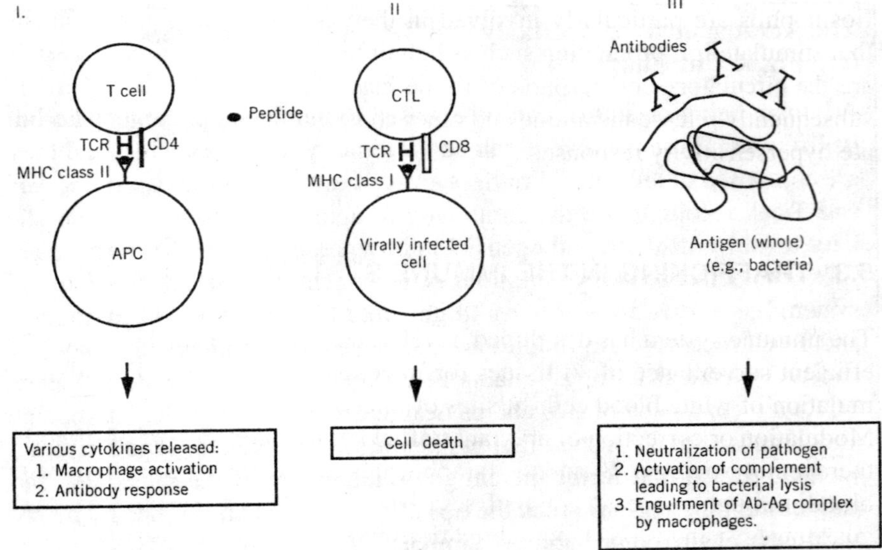

Figure 5.1. Important interactions in a normal immune response. (I) Helper T cells recognize peptides presented by MHC class II molecules on professional antigen presenting cells (APCs) via the TCR (H). This interaction leads to cytokine release and subsequent immune activation of T and B cells. These T cells carry a CD4 marker. (II) Cytotoxic T cells (CTLs) recognize peptides presented by MHC class I molecules on, for example, virally infected cells. These T cells carry a CD8 marker. CTLs kill the infected cells. (III) Antibodies (Y) recognize epitopes on the whole antigen.

other arms of the immune response and express a CD4 co-receptor and are thus called the helper T cells or CD4 T cells. These interactions are demonstrated in Fig. 5.1. Helper T cells can be further divided into two functionally distinct subsets: Th1 and Th2, which differ in their cytokine profiles and effector function (see Section 5.5). The primary function of the mononuclear phagocyte system is phagocytosis, a process by which an entire pathogen can be engulfed into the cell. Monocytes are blood cells that differentiate upon settling into particular tissues into macrophages (different tissue macrophages are assigned different names). They present antigen to helper T cells (and are therefore termed antigen presenting cells or APCS) and also capture antigens tagged by an antibody or a complement protein for disposal. Interdigitating dendritic cells are found in the interstitium of most organs and serve like the macrophages to process and present the antigen to helper T cells. Granulocytes form the final set of immune cells. Neutrophils respond rapidly to chemotactic stimuli and phagocytose foreign particles, and they are the major cell population in the acute inflammatory response.

Eosinophils are particularly involved in the destructive response to agents that stimulate IgE production such as helmintic parasites (worms). Basophils are the circulatory counterparts of tissue mast cells. Both bind free IgE and subsequently release histamine and other chemical mediators of the immediate hypersensitivity response.

5.3. TRAFFICKING IN THE IMMUNE SYSTEM

The immune system has developed novel trafficking mechanisms that enable efficient surveillance of all tissues for infectious pathogens and rapid accumulation of white blood cells at sites of infection or injury (Springer 1994). Modulation or correction of aberrant trafficking has proved to be an effective therapeutic strategy. B and T lymphocytes continually recirculate from blood, through tissue, into lymph and back to blood to provide a constant guard against invasion. Upon activation, lymphocytes are "tagged" with a receptor that allows them to return back to the site of their original interaction with antigen. Similarly monocytes and granulocytes emigrate from the bloodstream to the area of infection or injury in response to particular stimuli; they do not recirculate, however. The accumulation of monocytes and granulocytes at the required site is postulated to occur in three sequential, but overlapping, steps. Initially an interaction between so-called selectins and their cognate carbohydrate receptors, located on the surface of the leukocytes and the damaged endothelium, respectively, slows the progress of the circulating leukocyte. Subsequently inflammatory signals mediated by chemoattractants are received by chemoattractant receptors on the leukocytes. These signals may derive from remnants of an initial nonspecific immune response at the lesion, such as bacterial protein fragments or C5a, from complement or small cytokines known as chemokines which are also the corollary of injury or infection. This second signal then stimulates a firmer interaction between integrins or adhesins on the leukocytes with ligands on the endothelium designated as addressins. This signal serves to arrest the movement of the leukocytes. In the case of normal B and T lymphocytes, they circulate from blood to lymphoid organs using homing receptors that recognize lymph node and blood vessel addressins. Upon antigen activation, however, they lose these organ homing receptors and gain receptors that then allow them to enter the bloodstream to migrate to inflammatory sites as well as the liver, the lung, and the spleen. As a result numerous molecular therapeutics have been proposed against inflammation that target the selectins, chemokines, integrins, adhesins, and the addressins (see Section 5.7).

Communication between immune cells is mediated by soluble proteins called *interleukins* or *cytokines.* As with chemoattractants, these molecules exert their biological functions through specific receptors expressed on the surface of target cells. Binding of the receptors triggers the release of a cascade of biochemical signals which profoundly affect the behavior of the cell bearing that receptor. Many cytokines and their receptors have been identified at the molecular level (Paul and Sedar 1994) and make suitable molecules of therapeutic value, as well as therapeutic targets in their own right. A brief list of their effects is tabled below:

CYTOKINE	EFFECTS
IL-1	Mediator of inflammatory response to bacteria (cf. TNFα)
IL-2	T-cell growth factor (Th-1), NK and B cell growth function stimulated
IL-3	Promotes development of all myeloid cell types
IL-4	T-cell growth (Th-2), IgE production
IL-5	B-cell growth, Eosinophil activation
IL-6	B-cell growth stimulator, role in inflammatory response
IL-7	B-cell development
IL-9	Promotes myeloid cell development
IL-10	B-cell activation, Th-2 response
IL-11	Development of platelet precursors
IL-12	Th-1 stimulator, NK stimulator, CTL maturation
TNF-α	Anti-microbial inflammatory response, anti-tumor response
IFN-(α, β, γ)	Anti-viral response
TGF-β	Negative regulator of immune response

Most cytokines have multiple effects which they often share with other cytokines. This property, referred to as *pleitropy,* is explained by the molecular structures of their receptors, which are composed of multiprotein combinations often containing a common functional domain (in the form of a common protein subunit) and enabling different receptors to mediate a similar effect. Cytokines allow communication between all of the leukocytes. The T lymphocytes are the central regulatory cells of the immune system and mediate their own activation and that of B lymphocytes. Cytokines also serve to determine the progression of the immune response toward a cell-mediated Th1 response or an antibody-dominated Th2 response. Like the adhesins and addressins, other receptors exist on T and B lymphocytes to allow physical interaction and thus efficient communication between them. The immune system can thus be thought of as a highly integrated unit composed of circulating cells that sense and respond to foreign antigen at any location in the body. This unit is controlled by physical interactions and by protein intermediates that link the various cells.

5.4. ANTIGEN PROCESSING AND PRESENTATION BY THE IMMUNE SYSTEM

The immune system has evolved to specifically detect and eliminate foreign material within a host (Germain 1994). This material may be of a viral, bacterial, or parasitic origin and may reside outside or within the cells of the host. Even within the cell, the foreign material can reside in distinct subcellular organelles or compartments. The immune system has developed strategies to cope with all of these scenarios.

Immune recognition of foreign material is of two types—general and specific. The recognition of idiosyncratic repeating polysaccharides in the cell wall of bacteria is an example of a general mechanism used to deal with microorganisms. However, most immune responses consist of a specific recognition of foreign antigens. This recognition is mediated by the immunoglobulins (Igs) and the T-cell receptors (TCRs). Immunoglobulins or antibodies form the receptors and effectors (in membrane-bound or secreted forms) of B cells. They can interact with a broad range of chemical structures, but their ligands must generally be on the outside of the invader, secreted by this organism, or expressed intact on the surface of infected host cells. Intracellular antigens are sequestered or hidden from this B-cell arm of the immune response and are dealt with instead by T cells. Two pathways of antigen processing exist to convert antigens into MHC molecule-binding peptides (Fig. 5.3). The existence of this dichotomy allows the immune system to scan peptides from two different locales: so-called endogenous (within the cell) peptides from the cytosol and exogenous (outside the cell) peptides from endosomes. The MHC is a genetic region in which are encoded many products responsible for providing extracellular presentation of intracellular invasion.

The MHC Genes and Products

Early immunologists studying tissue transplantation in mice discovered a locus containing numerous genes that controlled graft rejection. This region was hence termed the *major histocompatibility complex* (MHC). The genetic elements controlling immune responses and determining the specificity of T-lymphocyte antigen recognition were also found to lie in the MHC locus. We now know that these properties are a result of the polymorphism and peptide presenting activities of the MHC molecules.

The human MHC is often referred to as the human leukocyte A (HLA)

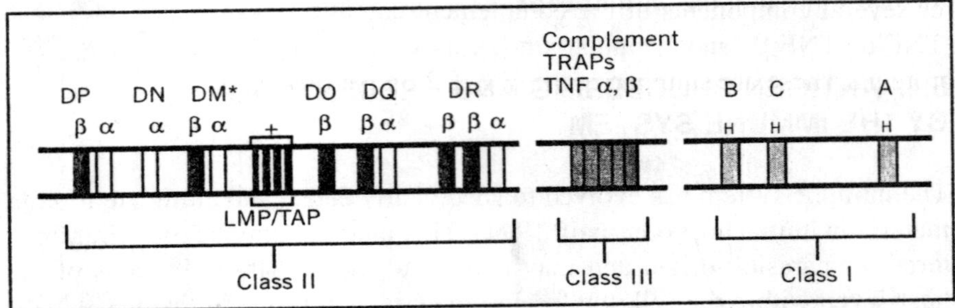

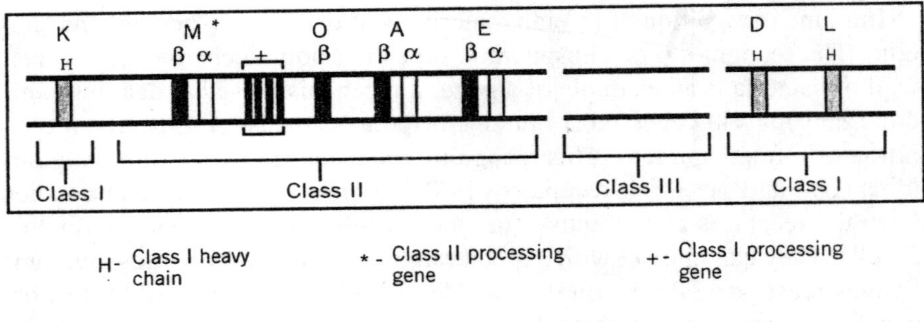

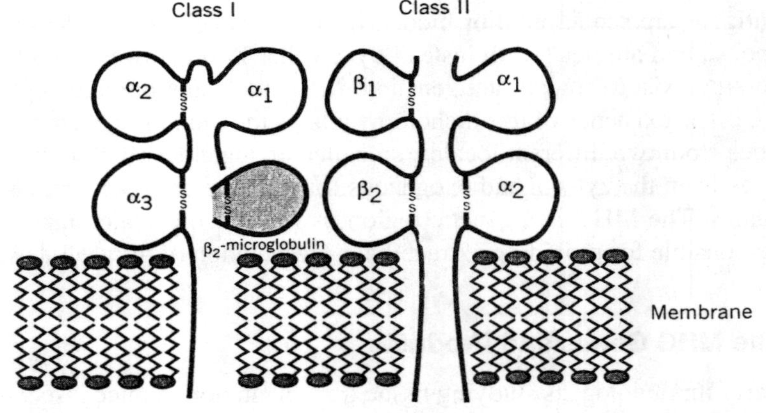

Figure 5.2. The organization of the MHC and structure of MHC molecules.

complex and the mouse MHC as H-2. Both can be divided into three functional regions. The class I region contains genes for three related proteins which, together with beta-microglobulin, form the class I molecule. The class II region encodes both chains of three heterodimeric class II molecules. In addition this region encodes a gene involved in the class II processing pathway (termed DM), and four components critical for class I processing

(LMP2, LMP7, TAP1, and TAP2). Finally, the class III region contains genes for several components of the complement cascade, some of the cytokines (TNF-α, TNF-β, and lymphotoxin-β), as well as other nonimmunological proteins. These are summarized in Fig. 5.2 on page 162.

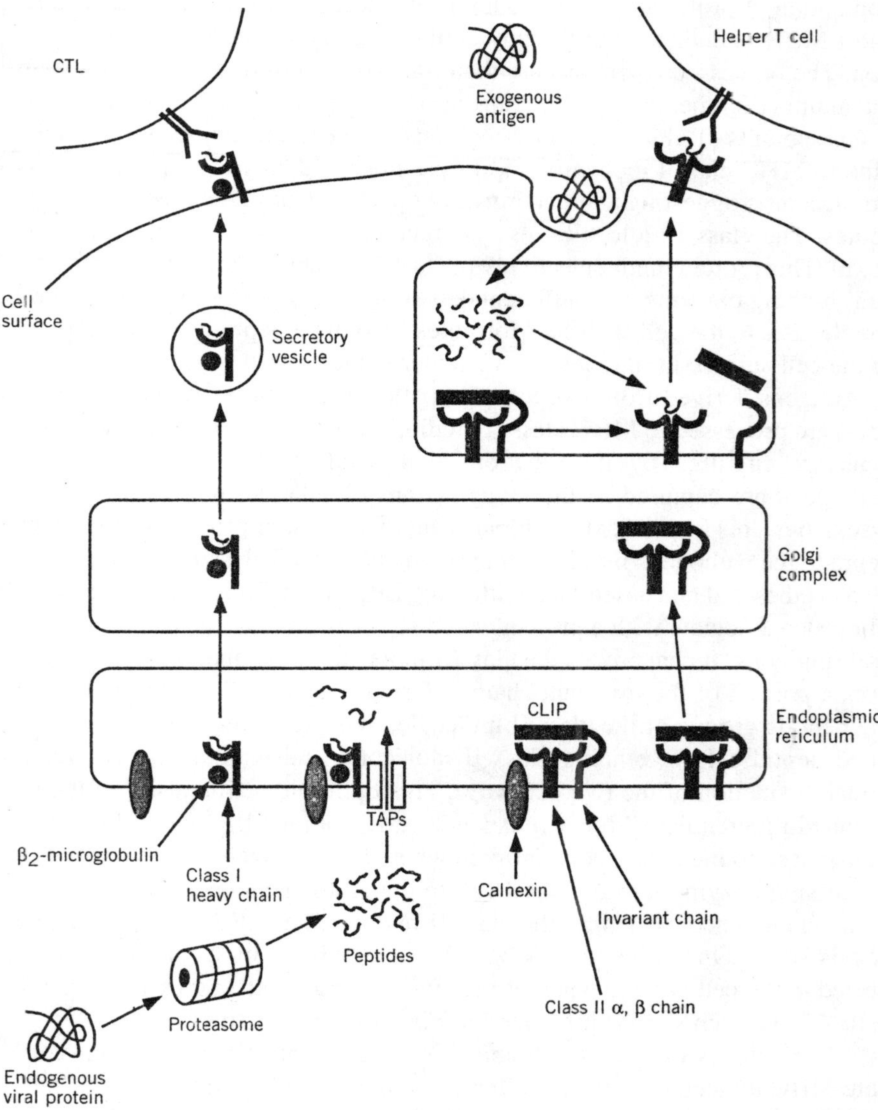

Figure 5.3. The pathways of antigen processing.

Processing Pathways

Cytosolic antigens, such as those of a viral origin, are postulated to be processed into smaller fragments by a multicatalytic, polyprotein complex referred to as proteasome that is found in all cells. These large cylindrical particles are normally responsible for the clearance of "useless" proteins within the cell. Such proteins may be misfolded or damaged, and they are tagged for clearance by the small protein ubiquitin. It is thought that a subpopulation of proteasomes that contains two MHC-encoded proteins (LMP2 and LMP7) is dedicated to the processing of antigens for immune recognition. The processed fragments are then transported through the endoplasmic reticulum (ER) membrane by a translocator composed of two transmembraneous subunits (TAPs). Once in the lumen of the ER, the antigenic fragments bind to MHC class I molecules. TAPs are noncovalently associated to class I molecules, suggesting a direct transfer of peptide from TAPS to class I molecules. The class I molecules also interact with another protein called calnexin. This protein binds selectively to class I molecules devoid of a peptide. This binding prevents the inefficient departure of "empty," functionless class I molecules to the cell surface. Molecules carrying a peptide are chaperoned to the cell surface in intracellular vesicular vehicles.

Antigens derived from extracellular pathogens, such as bacteria or parasites, are processed to MHC class II binding peptides. These exogenous antigens are engulfed by invaginations that form at the cell surface. The pathogens are captured as the invaginations circularize into endosomes or lysosomes. It is in this acidic compartment that a host of proteolytic enzymes degrade the pathogens into antigenic fragments. MHC class II molecules are also synthesized and assembled within the ER, like their class I counterparts. They also associate with a protein called the *invariant chain* that serves several functions. It chaperones the class II molecule to its final locale. In addition, a portion of the invariant chain, referred to as the CLIP region, blocks the peptide groove of the class II molecule, thus preventing any class I destined peptides from binding class II molecules and maintaining the functional distinction of the two pathways. This tight interaction also fulfills the "calnexin prerequisite" for permission to exit from the ER. This complex is transported to the acidic compartment where the pathogen is processed. The processing enzymes remove the CLIP to unveil the peptide-binding groove. A molecule, encoded within the class II region called DM, may play a role at this stage. Once a peptide is bound to the class II molecule, it is transported to the cell surface where it may then interact with surveying helper T cells. These events are summarized in Fig. 5.3 on page 163.

The outcomes of CD8 T-cell and CD4 T-cell interactions with their cognate MHC molecules are very different. In the case of the class I molecules, CD8 T-cell interaction and activation will trigger a reaction that ultimately leads to the destruction of the host cell by pore-forming perforins and other enzymes. Thus class I presentation is effectively a suicide signal. In the case of class II molecules, the antigen recognized has originated outside the cell and is sequestered in a compartment once inside. Thus the whole cell does not need to be sacrificed to eradicate the pathogen. Instead, MHC class II molecule interaction with CD4 T cells leads to a cascade that activates other components of the immune system toward the initiating pathogen.

Structure of MHC Molecules

MHC class I molecules are present on all nucleated cells of the body, although their levels of expression vary. In contrast, MHC class II molecules are expressed primarily by APCs such as B cells, macrophages, monocytes, and dendritic cells. MHC class I molecules are composed of a 45 kDa MHC-encoded heavy chain noncovalently associated with a 12kDa, non-MHC-encoded, soluble protein named *beta-microglobulin,* whereas class II molecules consist of a transmembraneous heterodimer of two MHC-encoded proteins (α-33kda and β-28kda). Despite these differences X-ray crystallographic analysis has shown the MHC structures to be remarkably similar.

In the class I molecule structure (Bjorkman et al. 1987), two immunoglobulinlike domains, corresponding to the membrane-proximal portions of the class I heavy chain and to beta-microglobulin, support a peptide-binding site composed of a floor of 8 beta strands flanked by helical walls, all from the heavy chain. The class II molecule structure contains a similar peptide-binding groove formed by a contribution of 4 beta strands and 1 helix from each of the 2 chains. The ends of the class I molecule groove are closed by other residues restricting the peptides bound to optimal lengths of 9–10 amino acids. In contrast, class II molecules form bonds with the main chain atoms of the peptide, and an open-ended groove allows them to bind peptides ranging from 12–24 residues.

In both cases the polymorphism characteristic of MHC molecules maps largely to the peptide-binding groove. Many regions of the human genome show polymorphism (differences of the same region in individuals from the same population). The MHC region is, however, one of the most polymorphic, and thus many MHC alleles exist. In addition, whereas most of the polymorphism of other parts of the genome lies in noncoding regions, the differences in MHC genes lie in their coding regions, and more specifically in the parts of the protein that bind peptide. This polymorphism has clearly defined functional effects. The recent X-ray crystallographic analysis of several MHC molecules, with and without peptide, have complemented the data obtained from the direct elution and identification of MHC-bound peptides to provide extensive detail of the interaction of different MHC alleles with their bound peptides.

Each MHC molecule can bind numerous peptides sharing a common "low-stringency" motif. This "broadness" of binding, together with the presence of several different MHC molecules, allows each individual's immune system to present a variety of pathogenic peptide antigens, and thus ensures that the individual has a greater chance of coping with any pathogenic assault. This information is being used to develop peptide and peptidomimetic agonists and antagonists to MHC molecules (Section 5.6).

5.5 MECHANISMS OF TOLERANCE INDUCTION

The mechanisms for T-cell tolerance are better defined than those for B cells. The general concepts described below apply mostly to both populations. One of the approaches the immune system has chosen is to assume that every protein found in the thymus during early development is derived from self-proteins, since the individual has yet to be exposed to external pathogens at this stage. Consequently T cells that recognize self-proteins are destroyed in the thymus, while those that do have potential to detect foreign proteins in the future are allowed to live. This is the premise behind central tolerance (Fig. 5.4).

Central Tolerance

Immature T cells migrate from the bone marrow to the thymus. Here they differentiate into mature CD4 and CD8 T cells, at the same time going through a process referred to as *thymic education.* T cells expressing self-reactive TCR are deleted upon engagement with a self-antigen and an MHC, a process known as *negative selection.* Those recognizing self-MHC only are positively selected. However, many self-antigens are not present in the thymus, making negative selection an incomplete screening process for self-reactive T cells. Mechanisms operating in the periphery (organs other than the thymus) compensate for this.

Peripheral Tolerance

Several mechanisms exist to inactivate self-reactive T cells in the periphery: clonal anergy, clonal deletion, clonal ignorance, and other forms of regulation/suppression. The "anergization" of a T cell results in its functional paralysis. This is postulated to occur in a series of steps: T-cell interaction with APCs normally occurs via accessory molecules present on both cell types (signal 1) and the TCR with the MHC (signal 2; see Fig. 5.4). One example of accessory protein pairing is that of the B cell B7 with with its cognate ligand CD28 on T cells. Both TCR and CD28 derived signals are capable of

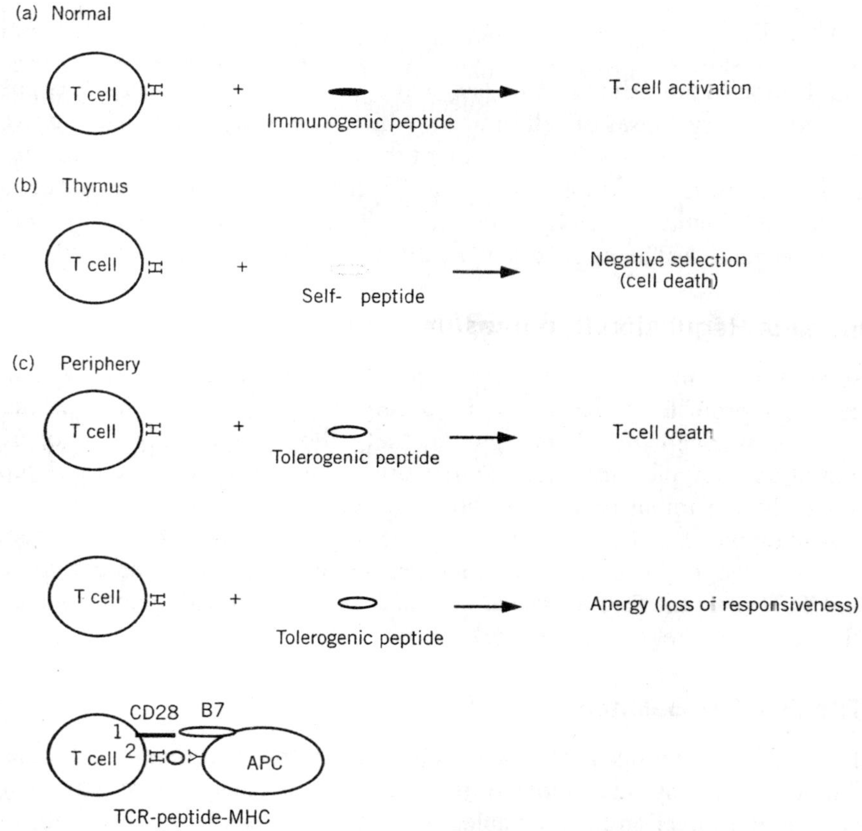

Figure 5.4. Mechanisms of immune tolerance. The TCR is represented by an H. (I) In a normal response a T cell will be activated by an immunogenic peptide presented by MHC molecules. (II) In the thymus during T-cell development, all T cells recognizing self-peptide presented by self-MHC will be deleted. (III) In the periphery a similar deletion may occur. This is controlled by various factors including the interaction of B7 and CD28 (see Section 5.5). Alternatively, the self-reactive T cells may be inactivated by various means. The two-signal hypothesis is represented in the bottom line. Signal 1 is through the B7-CD28 interaction, and signal 2 is through the TCR-MHC interaction. One signal leads to inactivation of the T cells, whereas two signals leads to activation.

modulating the function of the T cells. A combination of signals 1 and 2 (which would occur if B7 was induced on B cells during an infection) would lead to T-cell activation. However, in the absence of infection, a lack of B7 would result in only signal 2. In this case the T cell would be signaled to become unresponsive. Some evidence also exists for the downregulation of TCR, CD4, CD8, and other accessory molecules on self-reactive T cells.

Negative thymic selection occurs by a deletion mechanism in which self-reactive T cells are triggered to self-destroy by a process called *apoptosis.* Eradication of autoreactive T cells can also occur in the periphery. It is postulated that large doses of self-antigen, which lead to repeated (self-) activation of T cells may cause the deletion of those cells. Many self-antigens may also be sequestered within cells, be inefficiently processed, or be inaccessible to the immune system by other means. T cells recognizing such proteins are referred to as "being ignorant" of them until they become exposed.

Immune Regulation/Suppression

Several mechanisms have been proposed for immune regulation, although definitive proof for some models is lacking. The proposal for the existence of suppressor cells that downregulate other cells is such a case. The operation of an idiotypic–anti-idiotypic network is another. In the latter, it is proposed that an initial immune response is followed by the development of anti-idiotypic B and T cells that downregulate the primary response themselves by targeting the idiotype (antigen-binding region) of the antibodies and TCR from the first response. This may be followed by several such cycles to create a regulatory network.

The Th1-Th2 Balance

T and B cells may also influence each other by the secretion of cytokines. Studies analyzing the immune response of different mouse strains to *Leishmania* infection have revealed a dramatic difference in susceptibility, which has proved to be a result of a dichotomy in the helper T-cell response (Heinzel et al 1989). In resistant mouse strains, viable *Leishmania* stimulated a so-called Th1 response that cleared the pathogen by activating nitric oxide (a mediator of pathogen destruction) production in infected macrophages. In contrast, in susceptible strains, Th2 cells were induced that exacerbated disease by suppressing macrophage activation through IL-4 secretion. These cells derive from the same helper T-cell precursors, and this polarization to Th1 or Th2 is mediated through cytokines that lend selective effector functions to the cell types and also inhibit the response of the T cells of reciprocal type (Mossman and Coffman 1989).

	Cytokines Produced	Effector Function
Th1	IL-2	Activation of CTLs, NK cells Expansion of T cells
	γIFN	Cell-mediated immunity Antimicrobial

Th2	IL-4	IgE production (allergy)
	IL-5	Eosinophilia (antiparasite)
	IL-10	
	TGF-ß	(See oral tolerance)

The Th1-Th2 balance thus represents a major focus of immune regulation and therapeutic intervention. For example, pushing the response from Th2 to Th1 would reduce IgE production. This is one of the main components of histamine release and its consequent effects on allergic reactions.

5.6. SPECIFIC IMMUNE MODULATION

Tolerance Induction with Antigen

It has been long known that certain forms of antigen administration not only fail to elicit a T-cell immune response but render the host unable to respond to challenge with an immunogenic form of the same antigen; the host is said to be tolerized to that antigen. Three factors influence whether a foreign antigen induces immunity or tolerance in adults: the degree of antigen aggregation, the presence of an adjuvant (an agent capable of stimulating the immune response), and the route of administration (intravenous, oral, intradermal, etc.). Aggregated antigens induce immunity, although monomeric antigen can induce immunity if injected locally in an adjuvant like complete Freund's adjuvant (mineral oil containing heat-killed *Mycobacterium tuberculosis*). The presence of adjuvant is necessary for cell-mediated immunity if the antigen is monomeric and introduced via routes that result in systemic (whole body) delivery of antigen (intraperitoneal, intravenous). Thus administration of a protein can lead to a response or actually downregulate a response to itself, depending on the form and route of administration. This then provides a novel therapeutic agent—the self-protein.

In IDDM a panel of autoantigens (the focus of the aberrent self-response) has been identified (see sidebar on IDDM). Spontaneous T-cell responses are detected specifically in NOD mice and not in resistant strains, as would be expected, at a very early age (Tisch et al. 1993). One of the earliest responses is to GAD2, suggesting this to be a key target in eliciting disease and therefore a potential immunomodulator. Indeed, intrathymic injections into three-week-old NOD mice resulted in a marked reduction of T-cell proliferation to GAD2 and also prevented spontaneous development of IDDM. Similar injections with another candidate autoantigen failed to give the same result, indicating an antigen-specific intervention. However, this intervention was not due to thymic deletion, an the tolerance observed was actually widespread to

INSULIN-DEPENDENT DIABETES MELLITUS

Insulin-dependent diabetes mellitus (IDDM, also called juvenile diabetes or type 1 diabetes) afflicts 0.2% of the U.S. population, with an average age of onset of 11 to 12 years. The disease stems from an autoimmune destruction of the insulin-producing beta cells of the islet of Langerhans in the pancreas. The lack of insulin results in disregulated glucose homeostasis (uncontrolled levels of glucose in the blood) and subsequently ketoacidosis, polydipsia (thirst), and increased urine production. Continuous hormone replacement therapy is required, and despite this, long-term patients often suffer from kidney failure (due to glomerular injury), blindness, or gangrene. Two animal models of IDDM exist, the so-called BB rat and the nonobese diabetic (NOD) mouse. The latter shows remarkable similarity to the human condition and remains the best animal model for a spontaneous autoimmune disease. Studies with the NOD mouse show that overt diabetes is preceded by an infiltration of lymphocytes into the islets (referred to as *insulitis*). These infiltrates consist of CD4 helper T cells and CD8 cytotoxic T lymphocytes, although adoptive transfer experiments suggest CD4 T cells are of primary importance in targeting specifically the beta cells for destruction. The role of CD8 T cells, toxic nitric oxide free radicals (NO−) and cytokines such as TNF-α and IL-1 in this process have all been proposed but remain unclear. Certain microbial or viral agents, especially coxsackievirus B-4, have also been suggested to trigger the autoimmune attack, but again evidence for this remains inconclusive. Several autoantigen candidates have been recently identified. They include glutamic acid decarboxylase isoforms 1 and 2 (GAD1 and GAD2), carboxypeptidase H, peripherin, heat shock protein, PM1 (or P69), and preproinsulin and I-A2.

Of these, T-cell reactivity is directed temporally to GAD2 first in NOD mice. Genetic analysis of the NOD mouse has shown the existence of at least 10 loci involved in eliciting diabetes. Of prime importance is the major histocompatibility (MHC) region. Several genes in this region have been implicated in conferring susceptibility or resistance to disease. Interestingly alleles conferring susceptibilty or resistance vary only by a few or single amino acids. The changes lie in the peptide-binding groove of the MHC molecules. How these subtle changes affect the course of the disease is discussed in the main text.

Experimental Allergic Encephalomyelitis: A Murine Model for Multiple Sclerosis

Experimental allergic encephalomyelitis (EAE) in mice, rats, and guinea pigs represents another murine model for an autoimmune disease. EAE is induced by immunization of susceptible strains with myelin basic protein (MBP) or proteolipid protein (PLP) in complete Freund's adjuvant (CFA). Coadministration of pertussis toxin, which is proposed to disrupt the blood-brain barrier, is a requisite for disease induction. Within one to two weeks the animals develop encephalomyelitis characterized by perivascular infiltrates composed of lymphocytes and macrophages associated with demyelination in the brain and spinal cord. The consequences range from mild to total limb paralysis and can be ultimately fatal. In its chronic stage the disease resembles multiple sclerosis (MS). However, EAE is an induced and not spontaneous disease, and the autoantigen(s) involved in eliciting MS remain undefined. Despite this, EAE remains the best characterized murine model of an autoimmune disease. It is mediated by a CD4 T-cell population bearing an oligoclonal (limited number) T-cell receptor repertoire. Most of these cells are of the Th1 type. In addition the epitopes recognized by these encephalogenic T cells have been identified, and the individual residues within the epitopes required for binding MHC and TCR molecules defined. EAE has been utilized extensively as a model to test novel strategies attempting to reengineer the immune system and thereby prevent autoimmunity to this crippling disease.

other antigens, pointing to a general tolerance or suppression mechanism. To this end, antibody responses specific for GAD and the other autoantigens persist, suggesting a Th1 to Th2 switch. Thus knowledge of the autoantigen allows the manipulation of the immune response by simple injection to prevent a complex disease like diabetes. Intrathymic injections of young children are obviously not feasible, although injection of GAD2 intravenously has resulted in protection from IDDM. This treatment would not be useful in humans, since it relies on prediabetic prognosis, impossible at this time. A more useful therapy would prevent an ongoing disease. Furthermore treatment of pre-diabetic NOD mice with an autoantigen-specific peptide named HSP-p277 (Elias and Cohen 1995) has been shown to downregulate the autoimmune response to this peptide and prevent the further development of IDDM by unknown mechanisms.

In the EAE model it has been demonstrated that multiple injections of a high dose of MBP peptide results in apoptosis (programmed cell death) of encephalogenic T cells and the subsequent amelioration of clinical and pathogenic signs of EAE. Here the mechanism is obviously different from the GAD-treated mice. A mechanism termed *propriocidal regulation,* where T cells are highly sensitive to high doses of antigen, may be operational here; this sensitivity may represent a means to keep the normal immune response "in check" from hyperresponsiveness. This modality also would be difficult to administer in human beings.

Another means to induce tolerance has been the oral or nasal administration of antigen. Studies have demonstrated that the feeding of whole proteins or specific peptides can delay or suppress the pathogenic process in EAE and IDDM in NOD mice, as well as in experimental autoimmune uveoretinitis and adjuvant arthritis (two other animal disease models). Oral administration (feeding) represents the most feasible means to induce tolerance in humans. Indeed, administration of type II collagen (a putative autoantigen) to 28 patients with severe rheumatoid arthritis (RA) resulted in a minimal decrease in the number of swollen joints and, in four cases, complete remission of the disease (Trentham et al. 1993). It is postulated that a corollary to the oral introduction of antigen is the induction of CD8 antigen-specific T cells that produce TGF-β, an immunosuppressive cytokine. TGF-β is present in amniotic, cerebrospinal, and ocular fluids and in the gut and is responsible for the "immune deviation" characteristic of the T-cell response in these privileged sites. In addition TGF-β can directly affect T cells by downregulating integrins that promote homing to the brain or inflammed tissues, and upregulating integrins responsible for homing to the gut. Thus these CD8 regulatory cells are proposed to promote "antigen-driven bystander suppression" through the cytokines they secrete. This represents another advantage of oral administration; the antigen eliciting the disease does *not* need to be known. In addition an *ongoing* response is downregulated by oral tolerization. Before this type of immune reengineering becomes feasible in humans, the pathway(s) that induce regulatory T cells must be defined. Furthermore the effect of the genetic variation of the MHC and other immune-related genes on these pathways must be elucidated. Only then will universal protocols and regimens become applicable to humans.

MHC Blocking Strategies

The interaction of helper TCR with MHC class II molecules is the hub of the wheel of the immune response, since the corollary to this interaction is a series of steps that serve to activate B cells, macrophages, and cytokines whose release triggers widespread effects. Studies have shown the association of several autoimmune diseases with certain MHC class II alleles (see Section

5.4). It is proposed that these MHC alleles selectively present particular autoantigenic peptides that activate the autoimmune response only in susceptible individuals. Thus a specific target for immune intervention or modulation is the MHC-peptide-TCR complex. Antibodies against MHC molecules represent one approach to interfering with this interaction. The use of monoclonal antibodies represents a novel class of prophylactic therapeutics (see Section 4.2). Since antibodies are specific and their repertoire virtually limitless, monoclonal antibodies can be viewed as a natural blocking drugs. The main problem faced with using monoclonal antibodies in this way is that an immune response may occur against them, since most are made in mice (see Section 4.2 for a discussion of the human anti-mouse antibody response, or HAMA). Administration of anti–class II antibodies has proved an effective therapy for murine IDDM, EAE, and the autoimmunity seen in NZB/NZW F1 mice.

Another approach is the design of MHC blocking peptides. This strategy relies on the fact that MHC molecules are promiscuous receptors capable of binding many peptides and that our increasing knowledge of these peptides bound by the MHC allows us to design peptides that serve a blocking function. These facts can be used to design MHC blocking peptides that would compete for binding with and therefore antagonize the autoantigenic peptide.

The seminal studies to test this rationale were carried out in 1987 using *in vitro* experiments with soluble MHC molecules and T-cell hybridomas restricted by the same molecules. These studies confirmed the validity of a blocking approach and showed that peptides with a higher affinity inhibited better. However, these studies also underlined some potential problems of this approach. For example, only a small fraction of the MHC molecules bound the peptides. Second, it was possible that some competitors were themselves immunogenic, creating the possibility that they would enhance the immune response and exacerbate disease. Thus the search continues for higher-affinity nonimmunogenic MHC antagonists. Preferentially these should be self-peptides to alleviate the potential of autoimmunity. They should also bind a single allele. Finally, the mode of delivery and susceptibility of the peptide to cellular proteases are factors that must be considered as a guide for therapeutic efficacy of the peptide.

The effectiveness of peptide therapy has been studied in NOD mice and particularly in the EAE model. In the latter case it is known that the immunogenic epitopes presented by I-A^u (a designation for a particular mouse MHC class II molecule found in EAE susceptible strains) myelin basic protein (MBP) amino acids Ac1-11 (the "Ac" indicates an acetyl group found at the end of the peptide), while the dominant epitope presented by I-E^u (a different class II allele) MBP amino acids 35–47. the induction of EAE by immunization with MBP can be inhibited following co-injection with non-self, nonimmunogenic ovalbumin peptide amino acids 323–339 or through the use of synthetic peptide analogs of Ac1-11 (Gautam et al. 1992). Here, however, other mechanisms may be operational. The ovalbumin 323–339 peptide was the first example of the prevention of an autoimmune disease by an MHC blocking peptide that is unrelated to the disease-causing peptide and is itself nonimmunogenic. This peptide is theoretically the ideal MHC competitor in this strain.

Nevertheless, EAE is not an ideal model to study the potential of peptide therapy. Unlike most autoimmune diseases, EAE is a harsh disease of rapid onset and short duration. To study peptide therapy in an ongoing disease, the NOD mouse is being utilized. The NOD mouse has a unique MHC class II molecule not found in other nonsusceptible strains, designated I-A^{nod}. The beta chain of this molecule has a Serine residue at position 57 that is different from the Aspartic acid residue found in other strains. Interestingly this is similar to the finding that the presence of a noncharged residue at position 57 of the DQ (the human equivalent of mouse I-A) beta chain in humans is strongly associated with IDDM susceptibility. The amino acid at position 57 lies in the MHC peptide-binding groove forming a salt bridge at one end. It is probably critical for binding, and its presence may determine the binding of autoantigenic peptides. A search is under way for an I-A^{nod} competitor.

Using a synthetic peptide shown to inhibit antigen presentation to an I-A^{nod}-restricted T-cell hybridoma to treat prediabetic NOD mice, it has been possible to reduce the incidence of spontaneous diabetes. However, this peptide may have been antigenic, since antibody and T-cell responses were induced to the peptide. In addition continuous treatment was required for the animals to remain disease free. Vaysburd et al (1995) have also identified a peptide that binds to I-A^{nod} and prevents the transfer of disease; diabetogenic cells (isolated from the spleens of diseased NOD mice) are transferred to irradiated (therefore lacking any B or T cells) young NOD mice that have not yet developed diabetes. Normally disease is detected by 20–30 days in 80–90% of the animals. This was reduced by 20% if 1–2 mg of peptide was injected concomitantly with the transfer cells.

These studies are encouraging, but they also serve to highlight the problems of peptide-blocking strategies. Nevertheless, the future may be more promising. The isolation of MHC-bound peptides has led to the identification of MHC allele motifs and, together with the crystal structures of MHC molecules, has allowed a high-resolution map of the peptide-binding zone to be defined. This presents a perfect arena to test the technologies outlined in Chapters 3 and 4 and to develop small molecule drugs that will act as MHC-binding, competitor antagonists without inducing antigenic responses. The desired pharmacologic and biochemical properties can be incorporated into the drug design. The identification of autoantigens and the mapping of the precise epitopes should provide the framework for the drug design.

TCR Blocking Strategies

The other half of the MHC-peptide-TCR complex interaction that lends itself to intervention is the TCR. Although the TCR is, like the MHC, a polymorphic dimer, it is formed by very similar mechanisms to those that generate antibody diversity, and thus, unlike the MHC, in order of 10^{10} different TCRs can potentially be produced by each individual. It is understandable then that preliminary blocking strategies focused on the MHC. However, the TCR became a *bona fide* target for specific intervention with the discovery that many autoimmune disorders were caused by T cells utilizing a limited repertoire of TCR (Acha-Orbea et al. 1988).

Analysis of the encephalogenic T-cell clones from EAE afflicted PL/J mice, for example, showed that all encephalogenic T cells share Vα3 and at least 80% Vβ8.2 expression (the TCR is divided into two chains α and β with the variable and active region designated by a number; thus Vα3 is simply a designation for a particular α chain of a TCR). In B10.PL mice, Vβ8.2 bearing TCR are expressed by about 60% of these T-cell clones. However, EAE, EAN (experimental allergic neuritis) and uvetis stand alone as model diseases showing this strikingly limited repertoire of TCR. Analyses of human autoimmune disorders, such as MS, IDDM, and Crohn's disease, and of spontaneous animal models have shown a wide heterogeneity, or conflicting results, as in RA and MS, with patient-to-patient variation. This may be partially due to the fact that EAE is a strikingly short disease, and thus the T cells obtained represent a selective early population. In spontaneous diseases, although a T-cell population with limited TCR usage may prime the disease, the T cells analyzed represent a late, mixed population composed of disease-causing, and nonpathogenic clones activated due to the general immune activation occurring after disease onset. In humans there is no way to differentiate pathogenic from nonpathogenic T cells, thus making TCR analysis difficult to interpret. Nevertheless, treatment of mice with a Vβ8 specific monoclonal antibody before disease induction has prevented EAE induction with encephalogenic T-cell clones, encephalogenic peptides, and whole MBP. Even reversal of the established disease was achieved with a single dose of a Vβ8 monoclonal antibody.

In RA the predominant use of Vβ14 was shown in T cells obtained from the joints of patients but not in the periphery of the same individual. This increase in Vβ14 in the joints was suggested to be due to the presence of a superantigen in the joints (since superantigens bind to particular V regions and activate very large numbers of T cells carrying that receptor). In this case then, as in EAE, the TCR represents a viable target for specific immune intervention. However, for most spontaneous diseases considerable effort is being made to identify and isolate the earliest, disease-triggering T cells. Analysis of these clones should then make the TCR targeting route more viable.

T-Cell Vaccination

The use of activated, autoimmunogenic T-cell clones, rendered harmless by irradiation or other such means, to vaccinate against disease was first described by Cohen and colleagues (Mor and Cohen 1995). The rationale behind this approach was to induce a response to these T cells and thus elicit the idiotype–anti-idiotype network (see Section 5.5) to specifically down-regulate the disease-causing T cells. In animal models, such as adjuvant arthritis, EAE, and the NOD mouse, this method has been shown to be effective. In a related approach several groups have also vaccinated with peptides, whose sequences derive from the MHC peptide-binding region of the TCR of encephalogenic T cells, to successfully prevent EAE in animal models. The mechanism in both cases is uncertain, and so this modality remains to be fulfilled. In addition this approach requires detailed knowledge of the autoantigen, the availability of T cells directed against it, and definition of the TCR usage. These T-cell and TCR-related technqiues would be difficult to carry out in human beings, since they rely on isolation of T cells from individual patients. In addition the suppression mechanisms may operate systemically and thus affect the overall immune response detrimentally.

Tilting the Th1-Th2 Balance

The Th1-Th2 balance refers to the interconversion of the two different forms of helper T cells (see Section 5.4). The two forms have large-scale and opposing effects on the immune system. If an immune response favors Th1 cells, then these cells will drive a cellular response, whereas Th2 cells will drive an antibody-dominated response. The types of antibodies stimulated by these cells are also different; for example, the production of the IgE antibodies responsible for some allergic responses is induced by Th2 cells.

The reciprocal regulation of Th1 and Th2 provides an excellent opportunity for immune intervention. Use of IL-4 or IL-10 to drive Th1 cells to the Th2 lineage would suppress autoimmunity, whereas administration of IL-2 or IFN-γ would prevent allergic responses. Indeed, IL-4 administration has prevented diabetes in NOD mice. However, this balance is precarious, and minor modifications can potentially lead to gross changes. In addition, shifting from a Th1 to Th2 response may alleviate autoimmunity but also cause an allergic response. Methods to inhibit Th1 or Th2 selectively without driving the cells toward the reciprocal type need to be elucidated. For instance, Th1 cells could be tolerized for autoimmune disease prevention and Th2 for hypersensitivity responses.

5.7. NONSPECIFIC MODELS OF IMMUNE INTERVENTION

The design of specific modalities to intervene in aberrant immune responses requires prior knowledge of the (auto)antigens and specific MHC alleles involved, isolation of disease-specific T cells, and elucidation of the TCRs utilized by these clones. These data are not readily available for most immune-related disorders. Thus many nonspecific means of immunomodulation are currently being developed. In some cases partial specificity is obtained by targeting specific arms of the immune response or by the introduction of the agents involved at particular sites or timepoints.

Immunosuppressive Agents Targeting the T Cell

FK506 functions as an immunosuppressant through the inhibition of signal transduction pathways that elicit helper T-cell activation. This effect is mediated by the inhibition of calcineurin (an intermediary protein in helper T-cell activation) by a binary complex composed of FK506 and an immunophilin protein designated FKBP. Vertex Pharmaceuticals, Inc. is attempting to design small molecule antagonists of calcineurin inhibition that would perform the same role as FK506 to downregulate the immune response. This agonist would be useful in transplantation, autoimmunity, and allergic responses.

Cyclosporine (CsA) is a lipophillic cyclic undecapeptide produced by the fungus *Tolypocladium inflatum Gams.* Like FK506, CsA binds an immunophilin and inhibits antigen-mediated T-cell activation. During the last decade CsA has been the principle immunosuppressant employed in solid organ and bone-marrow transplantation. It is now emerging as a potential agent in the therapy of autoimmune disorders. CsA can ameliorate experimental autoimmune diseases including rat arthritis, rat uveitis, and EAE in mice. The continuous administration of CsA before the age of onset of disease resulted in prevention of diabetes in BB rats and in NOD mice. Patients suffering from RA, autoimmune uveitis, and psoriasis have shown improvements from CsA therapy. However, the efficacy of CsA in treating an ongoing spontaneous disease remains to be proved. This, the persistent side effects, and poor pharmacological properties of CsA prevent its wide use.

Mycophenolic Mofetil (MM) (SYNTEX, Palo Alto) is an enzyme inhibitor acting on the pathway of purine nucleotide biosynthesis. This drug can block lymphocyte proliferation and thus has potential in suppressing graft rejection and autoimmunity and inflammatory responses. Indeed, it has been reported to facilitate long-term allograft (heart) survival in rats. MM has also been used successfully to prevent the activation of autoimmune T cells; continuous treatment of disease-prone BB rats prevented both insulitis and diabetes.

In a related approach BioCryst Pharmaceuticals Inc. have utilized the X-ray crystallography-derived structure of purine nucleotide phophorylase to synthesize potent inhibitors to this enzyme (see Section 4.4 for details). They are currently utilizing one such product, designated BCX-34, in clinical trials to assess its efficacy as a therapy for psoriosis. The problem with immunosuppressive drugs such as cyclosporin and FK506 is that they target housekeeping proteins (proteins that perform a role that requires their constant presence in the cell) such as immunophilins. Thus new avenues are targeting tissue-specific transcription factors or specialized signal-transducing molecules to gain increased selectivity and minimize side effects.

The breakdown of inositol phosphate-glycan is a critical signal transduction step in the proliferative response elicited by IL-2. Cell Therapeutics, Inc. have developed small-molecule, synthetic antagonists that prevent this breakdown. Similarly oligonucleotides capable of inhibiting a key step in the IFN-γ induced signal transduction pathway have been defined by the Immunogenetics and Transplantation Laboratory of the University of California, San Francisco. IFN-γ stimulates inflammatory responses and augments graft injection due to its positive effect on class I expression. Both of these reagents have potential as immunosuppressants. Recently it has been demonstrated that patients lacking a functional T-cell protein-designated ZAP-70- show compromised T-cell function. ZAP-70 is a pivotal protein in transmitting the TCR signal in response to antigen recognition to downstream T-cell activating transcription factors. Pharmacological agents that can block ZAP-70-mediated catalysis should be effective against autoimmunity without the side effects of FK506 and CsA.

Targeting Accessory Molecules

Both T and B lymphocytes are studded with accessory proteins on their cell surface without which they show inefficient function. These include CD3,

CD4, B7, and CD40-p39. CD3 and CD4 complement the function of the TCR. B7 (and B7-2) are found on APCs and deliver the "co-stimulatory" signal through CD28 required for T-cell proliferation and IL-2 production. T cells interact through p39 with a receptor on B cells-CD40- which then facilitates antibody class switching and secretion by those B cells. Thus these molecules are yet more potential targets for immune intervention.

Administration, under different regimens, of monoclonal antibodies against CD3, CD4, B7, TCR, and class II MHC molecules has led to prevention or treatment of IDDM in NOD mice. In addition anti-CD4 can lead to remission of EAE. Anti-CD3 treatment can also induce a long-term remission from overt diabetes (Chatenoud et al. 1994). Finally, p39 blockade has proved to be successful in treating antibody-mediated autoimmune arthritis.

All of these targets are ubiquitous proteins essential in the development of lymphocytes and for efficient functioning of mature cells. The exact mechanisms behind their successful usage remain elusive. For this reason, although the rationale of their targeting is sound, their utility remains experimental.

Targeting Cell Adhesion Molecules

The recruitment of both autoimmune and inflammatory cells to their site of action requires homing receptors and addressins. Monoclonal antibodies against $\alpha 4$ or L-selectin have been able to prevent diabetes in NOD mice in a disease transfer model (Yang et al. 1993) as have antibodies recognizing I-CAM and LFA-1 (Hasagawa et al. 1994). Treatment of neonatal mice can also induce long term prevention of spontaneous diabetes in NOD mice. Collagen-induced arthritis and EAE were also prevented by these means. Still the exact mechanisms remain unclear.

Since monoclonal antibodies are difficult to generate *en masse*, and may induce conformational changes in their targets that may in turn lead to side effects, many companies are seeking chemical agents that will block and thus inhibit accessory protein function. Tanabe Research Laboratories have designed peptide and nonpeptide inhibitors of integrin $\alpha 4\beta 1$. Alkermes, Inc. are utilizing peptides from a bacterial adhesion molecule in an attempt to inhibit inflammation. Texas Biotechnology Corp. and the Monsanto Company are utilizing phage display to isolate new antagonists of I-CAM, a common adhesion molecule. Since adhesion molecules contain critical glycosegments, glycomimetics are also being developed (Glycomed, Inc.). Parke-Davis, Tularik Inc., Cytel Corp., and Scios Nova, Inc. are developing similar small molecule antagonists (see Chapters 3 and 4) targeting adhesion and other immune molecules.

Targeting Cytokines and Related Molecules

Since cytokines have pleitropic effects, their modulation has potential use in a range of diseases including inflammatory and autoimmune disorders. The so-called inflammatory cytokines IL-1 and TNF-α play a prominent role in sepsis, inflammatory bowel disease, RA (Maini et al. 1995), and IDDM. They are the focus of many biotechnology ventures such as Immunex Corp. who have produced soluble ligand binding portions of IL-1 and TNF receptors to act as functional antagonists. Hoffman La Roche and Amgen are also developing similar agents. The efficacy of these soluble molecules is currently being tested in clinical trials with patients suffering from RA and other inflammatory diseases.

Antibodies against TNF-α have also been utilized as antagonists to prevent IDDM in NOD mice, and anti-TNF-α therapy has been shown to alleviate RA in humans (Panayi 1995). However, the precise mechanism here remains unclear. Another approach is to target proteins involved in the processing of these cytokines. Interleukin-1 converting enzyme (ICE) is a cysteine protease responsible for the conversion of inactive IL-1 precursor to its active form. The crystal structure of ICE coupled with mutational analysis has allowed the active site of the enzyme to be delineated. Both Merck, Sharpe & Dohme and Sterling Winthrop have developed ICE inhibitors. The latter have defined both peptide (acyloxy) methy ketones and nonpeptide agents through chemical file screening (see Chapter 4). Immunex Corp. has developed inhibitors of a metalloprotease that cleaves TNF-α to yield the soluble form of this cytokine. Similarly Lisofylline is a small molecule inhibitor of phosphatidic acid, a mediator of signal transduction by IL-8 and possibly other cytokines. It has been used to treat endotoxin shock and IL-8-mediated lung injuries in animal models. These modes of immune intervention provide a greater specificity, since agents blocking cytokine binding may act on related cytokine receptors.

5.8 SUMMARY

Currently available prophylactics for immunomodulation are conventional drugs such as CsA, FK506, cyclophosphamide, and azathioprine (Becker et al. 1995 lists such agents used for MS treatment) that are nonspecific. In addition some of these are toxic, and their continued use can lead to oversuppression of the immune system, which in turn facilitates the occurrence of lymphomas and viral infections. Better small molecule antagonists can now be created using the techniques decribed in Chapters 3 and 4.

The tools of modern biology have identified and made available a number

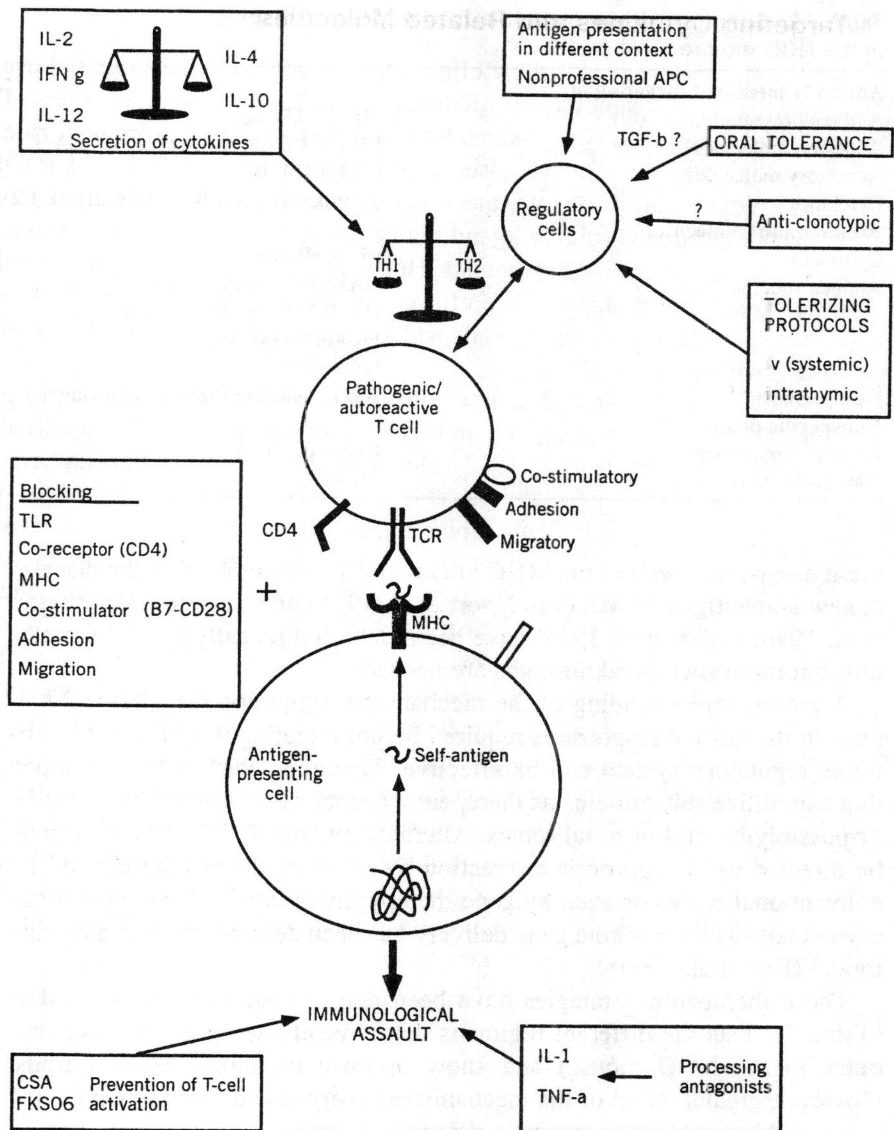

Figure 5.5. Strategies of immune intervention.

of targets for far more specific immune intervention. These are summarized in Fig. 5.5. The complexity of the immune system, however, requires the development of therapeutic agents that will provide greater specificity by inducing antigen-specific unresponsiveness to autoimmune disorders, for example. This will require identification of autoantigens involved in eliciting

Table 5.2. Summary of the methods used to prevent diabetes in the NOD mouse

Antibody mediated targeting of	
Antigen presentation	MHC, TCR
Adhesion molecules	LFA-1, I-CAM-1, L-selectin, VLA-4
Accessory molecules	CD3, CD4, B-7
Cytokines	TNF-α (Age-dependent), IL-1R, IL-12
Recombinant molecules	
Cytokines	rIL-4, rIL-10
Antigen specific	GAD65 (intravenous, intrathymic)
	GAD67 (footpad injections)
	Insulin (nasal, oral)
	HSP-p277
Nonspecific	BCG, complete Freund's adjuvant
Nonspecific drugs	
Immunosuppressant	Cyclosporine A
Anti-proliferative	Imuran

these diseases, as well as the MHC alleles and TCRs involved in the disease. A new autoantigen of MS (van Noort et al. 1995) and diabetes (Neophytou et al. 1996; Arden et al. 1996) have been identified recently by novel methods, but more such breakthroughs are needed.

A greater understanding of the mechanisms regulating the Th1-Th2 balance in the normal response is required before targeting this area of the immune regulatory system can be effective. Therapies need to be developed that can utilize self-proteins as therapeutic agents, administered by injection or possibly by oral or nasal routes. Alternatively, the Th1-Th2 balance may be directed to the appropriate direction by cytokine therapy introduced by conventional routes or even by gene therapy; modulation of systemic lupus erythematosus by cytokine gene delivery has been demonstrated in a murine model (Raz et al. 1995).

These therapeutic strategies have been demonstrated in animal models (Table 5.2 lists the different regimens that have elicited prevention of diabetes in the NOD mouse) and show promise in human clinical trials. However, greater detail of the mechanisms involved and the safety and efficacy of these molecules must be determined before such therapies can be used routinely in humans.

REFERENCES

Abbas, A. K., Lichtman, A. H., and Prober, J. S. 1995. *Cellular and Molecular Immunology,* 2nd ed. New York: Saunders.

Acha-Orbea, H., Mitchell, D. J., Timmermann, L., Wraith, D. C., Tausch, G. S., Waldor, M. K., Zamvil, S. S., McDevitt, H. O., and Steinman, L. 1988. Limited heterogeneity of T cell receptors from lymphocytes mediating autoimmune encephalomyelitis allows specific immune intervention. *Cell* 54:263–273.

Arden, S. D., Roep, B. O., Neophytou, P. I., Usac, E. F., Duinkerken, G., De Vries, R. R., and Hutton, J. C. 1996. Imogen 38: A novel 38-kD islet mitochondrial autoantigen recognized by T cells from a newly diagnosed type 1 diabetic patient. *Journal of Clinical Investigation* 97:551–561.

Becker, C. C., Gidal, B. E., and Fleming, J. O. 1995. Immunotherapy in multiple sclerosis. *American Journal of Health-Systems Pharmacy* 52:1985–2000.

Bjorkman, P. J., Saper, M. A., Samraoui, B., Bennett, W. S., Strominger, J. L., and Wiley, D. C. 1987. Structure of the human class I histocompatibility antigen. *Nature* 329:506.

Breedveld, F. C., Struyk, L., van Laar, J. M., Miltenberg, A. M., de Vries, R. R., and van den Elsen, P. J. 1995. Therapeutic regulation of T cells in rheumatoid arthritis. *Immunological Reviews* 144:5–16.

Bowman, M. A., Leiter, E. H., Atkinson, M. A. 1994. Prevention of diabetes in the NOD mouse: implications for therapeutic intervention in human disease. *Immunology Today* 15:115–120.

Chatenoud, L., Thervet, E., Primo, J., and Bach, J.-F. 1994. Anti-CD3 antibody induces long-term remission of overt autoimmunity in nonobese diabetic mice. *Proc. Natl. Acad. Sci.* 91:123.

Elias, D., and Cohn, I. R. 1995. Treatment of autoimmune diabetes and insulitis in NOD mice with heat shock protein 60 peptide p 277. *Diabetes* 44:1132–1138.

Gautam, A. M., Pearson, C. I., Sinha, A. A., Smilek, D. E., Steinman, L., and McDevitt, H. O. 1992. Inhibition of experimental autoimmune encephalomyelitis by a nonimmunogenic non-self peptide that binds to I-A^u. *Journal of Immunology* 148:3049.

Germain, R. N. 1994. MHC-dependent antigen processing and peptide presentation: Providing signals for T lymphocyte activation. *Cell* 76:287–299.

Hasegawa, Y., Yokono, K., Taki, T., Amano, K., Tominaga, Y., Yoneda, R., Yagi, N., Maeda, S., Yagita, H., and Okumura, K. 1994. Prevention of autoimmune insulin-dependent diabetes in non-obese diabetic mice by anti-LFA-1 and anti-ICAM-1 mAb. *International Immunology* 6:831–838.

Haskard, D. O. 1995. Cell adhesion molecules in rheumatoid arthritis. *Current Opinion in Rheumatology* 7:229–234.

Heinzel, F. P., Sadick, M. D., Holaday, B. J., Coffman, R. L., and Locksley, R. M. 1989. Reciprocal expression of interferon gamma or interleukin 4 during the resolution or progression of murine leishmaniasis. Evidence for expansion of distinct helper T cell subsets. *Journal of Experimental Medicine* 169:59–72.

Klareskog, L., Ronnelid, J., and Holm, G. 1995: Immunopathogenesis and immunotherapy in rheumatoid arthritis: an area in transition. *Journal of Internal Medicine* 238:191–206.

Krensky, A. M. 1995: HLA-derived peptides as novel immunosuppressives. *Pediatric Research* 38:275–279.

Lanzavecchia, A. 1993. Identifying strategies for immune intervention. *Science* 260:937–944.

Maini, R. N., Elliot, M. J., Brennan, F. M., and Feldmann, M. 1995. Beneficial effects of tumour necrosis factor-alpha (TNF-alpha) blockade in rheumatoid arthritis (RA). *Clinical and Experimental Immunology* 101:207–212.

Marmont, A. M., and Van Bekkum, D. W. 1995. Stem cell transplantation for severe autoimmune diseases: new proposals but still unanswered questions: *Bone Marrow Transplantation* 16:497–498.

Mor, F., and Cohen, I. R. 1995. Vaccines to prevent and treat autoimmune diseases. *International Archives of Allergy and Immunology* 108:345–349.

Mosmann, T. R., and Coffman, R. L. 1989. TH1 and TH2 cells: different patterns of lymphokine secretion lead to different functional properties. *Annual Reviews of Immunology* 7:145–173.

Neophytou, P. I., Muir, E. M., and Hutton, J. C. 1996: A subtractive cloning approach to the identification of mRNAs specifically expressed in pancreatic β cells. *Diabetes* 45:127–133.

Panayi, G. S. 1995. The pathogenesis of rheumatoid arthritis and the development of therapeutic strategies for the clinical investigation of biologics. *Agents and Actions* 47:1–21.

Paul, W. E., and Sedar, R. A. 1994. Lymphocyte response and cytokines. *Cell* 76: 241–251.

Pennline, K. J., Roque-Gaffney, E., and Monahan, M. 1994. Recombinant human IL-10 (rHUIL-10) prevents the onset of diabetes in the nonobese diabetic (NOD) mouse. *Clinical Immunology and Immunopathology* 71:169–175.

Rabinovitch, A. 1994. Immunoregulatory and cytokine imbalances in the pathogenesis of IDDM. Therapeutic intervention by immunostimulation? *Diabetes* 43:613 621.

Raz, E., Dudler, J., Lotz, M., Baird, S. M., Berry, C. C., Eisenberg, R. A., and Carson, D. A. 1995. Modulation of disease activity in murine systemic lupus erythematosus by cytokine gene delivery. *Lupus* 4:286–292.

Rizzo, L. V., and Caspi, R. R. 1995. Immunotolerance and prevention of ocular autoimmune disease. *Current Eye Research* 14:857–864.

Springer, T. A. 1994. Traffic signals for lymphocyte recirculation and leukocyte emigration: The multistep paradigm. *Cell* 76:301–304.

Theofilopoulos, A. N. 1995a. The basis of autoimmunity: Part I. Genetic predisposition. *Immunology Today* 16:90–98.

Theofilopoulos, A. N. 1995b. The basis of autoimmunity: Part II. Mechanisms of aberrant self-recognition. *Immunology Today* 16:150–159.

Tisch, R., Yang, X.-D., Singer, S. M., Liblau, R., Fugger, L., and McDevitt, H. O. 1993. Immune response to glutamic acid decarboxylase correlates with insulitis in non-obese diabetic mice. *Nature* 366:72–75.

Trentham, D. E., Dynesius-Trentham, R. A., Oray, E. J., Combitchi, D., Lorenzo, K. L., Sewell, K. L., Hafler, D., and Weiner, H. L. 1993. Effects of oral administration of type II collagen on rheumatoid arthritis. *Science* 261:1727.

van Noort, J. M., Sechel, A. C., Bajramovic, J. J., Ouagmiri, M. E., Polmann, C. H., Lassmann, H., and Ravids, R. 1995. The small heat-shock protein alpha β-crystallin as candidate autoantigen in multiple sclerosis. *Nature* 375:798–801.

Vaysburd, M., Lock, C., and McDevitt, H. 1995. Prevention of insulin-dependent diabetes mellitus in nonobese diabetic mice by immunogenic but not by tolerated peptides. *Journal of Experimental Medicine* 182:897–902.

Yang, X.-D., Karin, N., Tisch, R., Steinman, L., and McDevitt, H. O. 1993. Inhibition of insulitis and prevention of diabetes in nonobese diabetic mice by blocking L-selectin and very late antigen 4 adhesion receptors. *Proceedings of the National Academy of Sciences* 90:10494.

APPENDIX

BIOPHARMACEUTICAL COMPANY LIST

Abbott Laboratories, Inc.
1 Abbott Park Road
Abbott Park, IL 60064

Agouron Pharmaceuticals, Inc.
3565 General Atomics Court
San Diego, CA 92121

Agracetus, Inc.
8520 University Green
Middleton, WI 53562

Alanex Corp.
3550 General Atomics Court
San Diego, CA 92121

Allelix Biopharmaceuticals Inc.
6850 Goreway Drive
Mississuaga, Ontario L4V 1V7

Alexion Pharmaceuticals, Inc.
25 Science Park
Suite 360
New Haven, CT 06511

American Cynamid Corp.
P.O. Box 400
Clarksville Road
Princeton, NJ 08540

Amgen, Inc.
1840 De Havilland Drive
Thousand Oaks, CA 91320

Amylin Pharmaceuticals, Inc.
9373 Towne Center Drive
Suite 250
San Diego, CA 92121

Anergen, Inc.
301 Penobscot Drive
Redwood City, CA 94063

AntiCancer, Inc.
7917 Ostrow Street
San Diego, CA 92111

Aphton Corporation
P.O. Box 1049
Woodland, CA 95776

Ares Advanced Technology, Inc.
280 Pond Street
Randolph, MA 02368

Arqule, Inc.
200 Boston Avenue Suite 3600
Medford, MA 02155

Arris Pharmaceutical Corp.
385 Oyster Point Blvd. Suite 11
San Francisco, CA 94080

Beckman Instruments, Inc.
2500 Harbor Blvd.
Fullerton, CA 92634

Berlex Biosciences
15049 San Pablo Avenue
Richmond, CA 94804

BioCryst Pharmaceuticals, Inc.
2190 Parkway Lake Drive
Birmingham, AL 35244

Biogen, Inc.
Fourteen Cambridge Center
Cambridge, MA 02142

Bionumerik Pharmaceuticals, Inc.
8122 Datapoint Drive, Suite 1250
San Antonio, TX 78229

Bristol-Myers Squibb, Inc.
5 Research Parkway
P.O. Box 5100
Wallingford, CT 06492

Cambridge NeuroScience, Inc.
One Kendall Square, Building 700
Cambridge, MA 02139

Cantab Pharmaceuticals Research Ltd.
184 Cambridge Science Park
Milton Road
Cambridge, CB4 4GN
U.K.

Cell Therapeutics, Inc.
201 Elliot Avenue West
Suite 400
Seattle, WA 98119

Cell GeneSys, Inc.
344A Lakeside Drive
Foster City, CA 94404

CellPro, Inc.
22215 26th Avenue SE
Bothell, Washington 98021

Cephalon, Inc.
145 Brandywine Parkway
West Chester, PA 19380

Centocor, Inc.
244 Great Valley Parkway
Malvern, PA 19355

Chiron Corp.
4560 Horton Street
Emeryville, CA 94608

Creative Biomolecules, Inc.
45 South Street
Hopkinton, MA 01748

Cubist Pharmaceuticals, Inc.
24 Emily Street
Cambridge, MA 02139

CV Therapeutics, Inc.
3172 Porter Drive
Palo Alto, CA 94304

Cypros Pharmaceuticals, Inc.
2732 Luker Avenue
Carlsbad, CA 92028

Cytel Corporation
3525 Johns Hopkins Court
San Diego, CA 92121

Darwin Molecular Corp.
22025 Avenue SE
Bothell, WA 98021

DNAX Research Institute
901 California Avenue
Palo Alto, CA 94304

DuPont Merck Pharmaceutical Corp.
DuPont Experimental Station
Wilmington, DE 19880

G.D. Searle, Inc.
4901 Searle Parkway
Skokie, IL 60077

GeneMedicine, Inc.
8301 New Trails Drive
The Woodlands, TX 77281-4248

Geneic Sciences, Inc. (Titan Pharmaceuticals, Inc.)
400 Oyster Point Boulevard
Suite 315
South San Francisco, CA 94080

Gensia, Inc.
9390 Towne Center Drive
San Diego, CA 92121

Genzyme Corp.
One Kendall Square
Bldg. 1400
Cambridge, MA 02139

Genentech, Inc.
460 Pt. San Bruno Blvd.
San Francisco, CA 94080

Geron Corporation
200 Constitution Drive
Menlo Park CA 94025

Gilead Sciences, Inc.
353 Lakeside Drive
Foster City, CA 94404

Glaxo Research and Development Ltd.
891-995 Greenford Road
Greenford, Middlesex UB60HE
Great Britain

Glycomed, Inc.
860 Atlantic Avenue
Alameda, CA 94501

ICOS Corp.
22021 20th Avenue SE
Bothell, WA 98021

IDEC Pharmaceuticals Corp.
11099 North Torrey Pines Road #160
La Jolla, CA 92037

ImClone Systems, Inc.
180 Varick St.
New York, NY 10014

ImmunoTherapeutics, Inc.
3233 15th South Street
Fargo, ND 58104

Immunex Corp.
51 University Street
Seattle, WA 98101

Incyte Pharmaceuticals, Inc.
3330 Hillview Avenue
Palo Alto, CA 94304

The Immune Response Corp.
5935 Darwin Court
Carlsbad, CA 92008

ISIS Pharmaceuticals, Inc.
2292 Faraday Avenue
Carlsbad, CA 92002

Ligand Pharmaceuticals, Inc.
9393 Towne Center Drive
San Diego, CA 92121

Markek Biosciences Corp.
6480 Dobbin Road
Columbia, MD 21045

Merck, Sharpe & Dohme, Inc.
Research Laboratories
126 E. Lincoln Avenue
Rahway, NJ 07065

Microcide Pharmaceuticals, Inc.
850 Maude Avenue
Sunnyvale, CA 94043

Millennium Pharmaceuticals, Inc.
640 Memorial Drive
Cambridge, MA 02139

Molecular Research Institute
845 Page Mill Road
Palo Alto, CA 94304

Molecumetics, Inc.
2023 120th Avenue NE
Bellevue, WA 98005

Monsanto Company
700 Chesterfield Village Parkway
Chesterfield, MO 63198

Oncogene Science, Inc.
80 Rogers Street
Cambridge, MA 02142

Organon Teknica, Inc.
Molenstraat 110
NL-5340 BH-OSS
The Netherlands

Panlabs, Inc.
11804 North Creek Parkway South
Bothell, WA 98011

Pharmacopeia, Inc.
101 College Road East
Princeton, NJ 08540

PRIZM Pharmaceuticals, Inc.
11035 Roselle Street
San Diego, CA 92121

Protein Design Labs., Inc.
2375 Garcia Avenue
Mountain View, CA 94043

Ribogene, Inc.
21375 Cabot Boulevard
Hayward, CA 94545

Sandoz Pharmaceuticals Corp.
59 Route 10
East Hanover, NJ 07936

SangStat Medical Corporation
1505 Adams Drive
Menlo Park, CA 94025

Scios-Nova, Inc.
2450 Bayshore Parkway
Mountain View, CA 94043

Schering-Plough Research Institute
2015 Galloping Hill Road
Kenilworth, NJ 07033

Scripps Research Institute
10666 North Torrey Pines Road
La Jolla, CA 92037

SCRIPTGEN Pharmaceuticals, Inc.
200 Boston Avenue
Medford, MA 02155

Selectide Technologies, Inc.
1580 East Hanley Road
Tucson, AZ 85737

SIBIA, Inc.
505 South Coast Blvd.
La Jolla, CA 92037

Sequana Therapeutics, Inc.
11099 North Torrey Pines Road
La Jolla, CA 92037

Smith Kline Beecham Pharmaceuticals, Inc.
709 Swedeland Road
King of Prussia, PA 19406

Somatix Therapy
850 Marina Village Parkway
Alameda, CA 94501

Syntex Corp. (Roche Biosciences)
3401 Hillview Avenue
Palo Alto, CA 94304

Terrapin Technologies, Inc.
750-H Gateway Blvd.
San Francisco, CA 94080

Texas Biotechnology Corp.
7000 Fannin Suite 1920
Houston, TX 77030

Tripos Inc.
1699 South Hanley Road
St. Louis, MO 63144

Tularik, Inc.
270 E. Grand Avenue
San Francisco, CA 94080

Vertex Pharmaceuticals, Inc.
40 Allston Street
Cambridge, MA 02139

Viagene, Inc.
11075 Roselle Street
San Diego, CA 92121

Vical, Inc.
9373 Towne Center Drive
Suite 100
San Diego, CA 92121

XOMA Corp.
2910 7th Street
Berkeley, CA 94710

ZymoGenetics, Inc.
1201 Eastlake Avenue East
Seattle, WA 98102

Glossary

Active site The amino acid residues at the catalytic center of an enzyme. These residues provide the binding and activation energy needed to place the substrate into its transition state and bridge the energy barrier of the reaction undergoing catalysis.

Activator A regulatory molecule that binds to a controlling element and positively stimulates one or more structural genes.

Adenine A purine base found in DNA and RNA.

Adenosine triphosphate (ATP) A nucleoside molecule that provides a source of energy (stored in each of the three phosphate bonds) for metabolic processes requiring energy transfers. ATP is the principal energy-storing molecule of the cell.

Adjuvant A substance that nonspecifically activates the immune system. Many therapeutic molecules such as vaccines are provided with an adjuvant in order to enhance their effect.

Affinity A measure of the strength of binding of one molecule to another, such as of a ligand to a receptor or a substrate to an enzyme.

Alpha helix A helical structure formed by polypeptide chains that is one of the secondary structural characteristics of proteins.

Allele A given form of a gene that occupies a specific position or locus on a chromosome. Variant forms of genes occurring at the same locus are said to be allelic to one another.

Allelic exclusion The expression in plasma cells of antibody genes found in only one member of a pair of homologous chromosomes.

Allergen An antigen responsible for producing allergic reactions by inducing the production of the IgE immunoglobulin.

Allergy A general term describing the immune response to nonpathogenic antigens, leading to inflammation and other deleterious effects.

Allozymes Different molecular forms of an enzyme coded by different allelic forms of a gene.

Alu family A common set of dispersed DNA sequences found throughout the human genome; each is about 300 bases long, and the sequences are repeated at least 500,000 times. Alu sequences are speculated to have originated from viral RNA sequences.

Amber codon The codon UAG that signals termination of the nascent peptide chain.

Amino acid One of the 20 chemical building blocks that are joined by amide (peptide) linkages to form a polypeptide chain of a protein.

Amino group The–NH_2 chemical group, found in amino acids, with a basic nature.

Amino terminal domain The end of a polypeptide chain containing a free amino group.

Aminoacyl-tRNA synthetase One of a group of 20 enzymes that are required to activate an amino acid and link it to its specific tRNA during protein synthesis at the ribosome.

Anaphase Stage in nuclear division when the chromosomes move to opposite poles.

Anergy Clinically, the absence of an expected cell-mediated immune reaction in sensitized organisms caused by inactivated B and T cells. Also used to describe inactivated B or T cells.

Angstrom (Å) A unit of measurement equal to 10^{-10} m. Commonly used by crystallographers and modelers to measure interatomic distances.

Antibody (also known as immunoglobulin) A protein molecule formed by rearrangement of immunoglobulin gene segments in response to a specific antigen that can recognize and bind to it.

Antigen Any foreign molecule that stimulates an immune response in a vertebrate organism. Most antigens are proteins such as the surface proteins of foreign organisms.

Antigenic determinant A single antigenic site on a complex antigenic protein or particle (also known as an epitope). For protein antigens, this usually corresponds to a peptide of a few amino acids.

Antigen-presenting cells (APC) A variety of lymphoid cells that carry and present an antigen in a form (usually a cleaved peptide) that can stimulate T lymphocytes.

Antisense DNA or RNA composed of the complementary sequence to the target DNA/RNA. Also used to describe a therapeutic strategy that uses antisense DNA or RNA sequences to target specific gene DNA sequences or mRNA in order to bind and physically inhibit their expression.

Antiserum Serum from an animal that contains antibodies against a particular antigen.

Ascites The effusion of fluids into the peritoneal cavity caused by tumor cells. Hybridomas grown in the peritoneal cavity of mice produce ascites containing high concentrations of a selected monoclonal antibody.

Apoptosis A form of programmed or controlled cell death. Cell death during thymic selection and cytoxic T-cell activation occurs using apoptosis, a process whereby a cascade of signals leads eventually to the breakdown of the cellular DNA resulting in cell death.

Attenuated vaccine A vaccine composed of live infectious organisms (typically whole viruses) that exhibit low virulence or that have been prevented from multiplying in their hosts, such as by exposure to radiation.

Autoantibodies Antibodies that are erroneously formed against self-antigens.

Autoimmunity An immune response to tissues, cells, or molecular components of the host. The pathological result of this response is an autoimmune disease.

Autoradiography A method used to locate radioisotope-labeled materials that have been separated in gels or are present in blots. The location of the radio-labeled material is determined by overlaying the test material with a photographic film.

Autosomes All the chromosomes in the genetic complement (genome) except the sex chromosomes.

Avidity A measurement of the binding of an antiserum to a complex antigen. The avidity of the antiserum depends on the affinities of the various antibodies present in the antiserum to different epitopes on the antigen.

B lymphocytes (B cells) Lymphocytes derived from bone marrow that express surface antibodies, which act as receptors for specific antigens. B cells play a major role in the humoral branch of the antibody response.

Back propagation Form of neural network used to correlate structures with activities.

Balloon angioplasty Medical procedure in which a catheter with a balloon at the tip is used to clear or repair damaged blood vessels.

Basal factors Nuclear components that associate with chromosomes and enhance gene transcription.

Base pair A pair of nitrogenous bases (one purine and one pyrimidine), held together by hydrogen bonds, that forms the core of DNA and RNA.

Basophils A multinucleated leukocyte (white blood cell) with granules containing acid glycoproteins that can catalyze anaphylactic reactions and that also releases histamine and other mediators of the hypersensitivity response.

Bence-Jones proteins Free monoclonal antibody light chains found in the urine of patients with multiple myeloma.

Beta strand One strand of the protein secondary structure known as the *beta sheet.*

Beta sheet A three-dimensional arrangement taken up by polypeptide chains that consists of alternating strands linked by hydrogen bonds. The alternating strands together form a sheet that is frequently twisted. One of the secondary structural elements characteristic of proteins.

Bidirectional replication DNA replication in which two replication forks move in opposite directions away from the same origin of replication (O of R).

Biosurface The region on a biological protein, enzyme, or receptor that acts as a three-dimensional target to ligands or other small molecules.

Blotting The process of transferring DNA, RNA, or protein from a polyacrylamide or agarose gel to a nitrocellulose filter for further analysis (see Southern blotting).

Blunt-end (ligation) The attachment of DNA bases to a fragment of DNA that contains no overhang at either end and consequently no DNA bases available for hybridization (cf. sticky-end ligation).

Bone marrow Site of origination of stem cells which are the precursors of all blood and immune cells.

Bradykinin A kinin involved in inflammatory reactions, generated following tissue injury. Bradykinin is involved in vasodilation, smooth muscle contraction, and increased vascular permeability and is an important target for anti-inflammatory drugs.

Cyclic adenosine monophosphate (cAMP) A molecule, derived from ATP, that interacts with specific hormones in cells and forms a complex with CAP allowing transcription of some genes. Also a mediator of intracellular signal transduction pathways.

Calcineurin An intermediate in the T-cell activation pathway.

CAP (catabolite activator protein) A positive control protein that complexes with cAMP and binds to a specific region of a promoter and stimulates the binding of RNA polymerase and hence transcription of some genes.

Carboxyl group The–COOH functional group, acidic in nature, found in all amino acids.

Carcinogen A chemical substance or physical agent that can effect a malignant transformation in a cell, causing it to become cancerous.

Carcinoma A tumor originating in epithelial cells such as the skin and lining of the digestive or respiratory tracts.

cDNA (copy DNA) A DNA strand copied from mRNA using reverse transcriptase. cDNA represents all of the expressed DNA in a cell.

cDNA library A set of DNA fragments prepared from the total mRNA obtained from a selected cell, tissue, or organism.

Complementary determining region (CDR) The hypervariable regions of an antibody molecule, consisting of three loops from the heavy chain and three from the light, that together form the antigen-binding site.

Cell cycle The process of cell division separated into four phases: G1, S, G2, and M, DNA replication is confined to the S phase, and cell division in the M (mitotic) phase.

Cell-mediated immunity The branch of the immune response that is mediated by the T lymphocytes.

Centimorgan A map unit or unit of physical distance on a chromosome. One centimorgan is equivalent to a 1% frequency of recombination between two linked genes.

Chaperones Term given to a group of proteins (chaperonins) that guides other proteins in their assembly and transport.

Chemotaxis The increased directional migration of cells in response to a chemical gradient.

Chimeras Animals that contain cells from two genetically different parents.

Chromatin The chromosome as it appears in its condensed state, composed of DNA and associated proteins (mainly histones).

Chromosome The structure in the cell nucleus that contains all of the cellular DNA together with a number of proteins that compact and package the DNA.

Cistron Another name for a gene, expressed as a unit of function. A segment of DNA that codes for a single polypeptide domain.

Class I, II molecules Proteins encoded by different clusters of genes within the major histocompatability complex (MHC; See MHC).

Class switching The change in a B-cell clone from the secretion of antibodies with the same V (variable) regions and heavy-chain C (constant) region to antibodies with a different heavy-chain C region and hence different class (e.g., IgE to IgG).

Clone A population of genetically identical cells.

Cloned DNA A collection of identical DNA fragments produced as a result of their replication after insertion into a suitable bacterial or viral vector system.

Cloning The formation of clones or exact genetic replicas.

Codon A sequence of three adjacent nucleotides that designates a specific amino acid or start/stop site for translation.

Complement A series of nine plasma proteins that are activated by antibody-antigen complexes resulting in a cascade leading to a destruction of the foreign substance and lysis of cellular elements.

Complementary determining region (CDR) The region on an antibody molecule, composed of six hypervariable beta-turn loops, that determines its specificity to a given antigen.

Conformation The precise three-dimensional arrangement of atoms and bonds in a molecule that describes its geometry and hence its molecular function.

Consensus sequence An idealized sequence that represents the nucleotides most often present at each position in a given segment of DNA, such as a promoter.

Constitutive synthesis Synthesis at an unchanging or constant rate regardless of a cell's requirements.

Contigs A fragment of DNA that has been sequenced as part of a shotgun sequencing project and will eventually be merged with other overlapping fragments of DNA to generate the finished sequence. A collection of contigs constitute a set of overlapping clones that together cover the target region to be sequenced.

Controlling element A genetic region, such as a promoter or operator, that can respond to an environmental signal and determine whether its associated structural gene should or should not undergo transcription.

Corepressor A small molecule that interacts with a repressor protein and allows it to combine with an operator and prevent transcription.

Copia A family of closely related DNA sequences in Drosophila capable of transposition that are known to code for very large amounts of DNA.

Cosmids Vectors that allow the insertion of long fragments, such as of human DNA (up to 50 kbases), and their subsequent expression and amplification in bacterial cells.

Covalent bond A strong interaction between atoms that share electrons such that each has filled its valence shells.

Crossreactivity The ability of an antibody or another molecule to react with (bind to) an antigen other than its cognate one. A measure of the relatedness of two different antigenic substances.

Crossing over An event that entails a reciprocal exchange of segments between two homologous chromosomes and results in the genetic recombination of linked alleles. Crossing over involves a breakage and repair mechanism acting on homologous DNA strands during meiosis.

Cyclophosphamide A cytotoxic drug frequently used as an immunosuppressant and anticancer drug.

Cyclosporin An immunosuppressive drug effective in preventing transplant rejection.

Cytokines Soluble protein molecules secreted by cells of the immune system that have far-ranging and powerful plieotropic effects on the immune response.

Cytoplasm The medium of the cell between the nucleus and the cell membrane.

Cytosine A pyrimidine base found in DNA and RNA.

Cytosol The fluid portion of the cytoplasm exclusive of the organelles.

Cytotoxic T cells A subpopulation of T cells that recognize peptides presented by MHC class I molecules. These cells exert their effect by recognizing and subsequently lysing the infected cells. CTLs are identified by their expression of the CD8 marker protein on their surfaces.

Deletion A chromosomal alteration in which a portion of the chromosome or the underlying DNA is lost.

Denaturation (of DNA) The separation of the two strands of DNA as a result of the disruption of the hydrogen bonds following exposure to high temperature or chemical treatment.

Dimer A molecule formed between two identical molecules thus having the same composition but twice the molecular weight of the original molecules.

Diploid A species having two complete haploid sets of chromosomes.

DNA (deoxyribonucleic acid) The chemical that forms the basis of the genetic material in virtually all organisms—composed of the four nitrogenous bases Adenine, Cytosine, Guanine, and Thymine—covalently bonded to form two repeating chains with a backbone of deoxyribose-phosphate that are hydrogen bonded as purine-pyrimidine pairs.

DNase (deoxyribonuclease) One of a series of enzymes that can digest DNA.

DNA fingerprinting A technique for identifying human individuals based on a restriction enzyme digest about tandemly repeated DNA sequences that are scattered about the human genome.

DNA ligase An enzyme that catalyzes phosphodiester linkages and repairs nicks in DNA strands.

DNA polymerase An enzyme that catalyzes the synthesis of DNA from a DNA template given the deoxyribonucleotide precursors.

DNA transfection The introduction of DNA into mammalian cells and its subsequent expression.

Domain A portion of a protein, usually a single polypeptide chain, that forms a single compactly folded, functional unit.

Dominant A gene that expresses itself in the presence of its allele. Any genetic trait that is expressed when present as a single allele.

Downstream The direction on a DNA template from the 3′ end to the 5′ end toward the site of the initiation of transcription.

Duplex Any molecule that is composed of two strands, such as double-stranded DNA.

Electrophoresis The use of an external electric field to separate molecules on the basis of their charge.

Endocytosis The process of capture and encapsulation of foreign particles by cells.

Endonuclease An enzyme that can break the phosphodiester bonds that occur internally within a DNA chain.

Endoplasmic reticulum (ER) A system of membranes distributed among the cytoplasm that provides sites for protein synthesis and intracellular transport.

Endosome An acidic compartment within the cell which is the site of interaction of MHC class II molecules with their cognate peptides.

Enhancers DNA sequences that can greatly increase the transcription rates of genes even though they may be far upstream or downstream from the promoter they stimulate.

Enzyme-linked immunosorbent assay (ELISA) An assay where an enzyme that catalyzes a colorimetric reaction is linked to an antibody, and the complex used to measure the amount of antibody bound to a hapten by measuring the intensity of the color produced.

Eosinophils Immune cells that are involved in the response to agents that stimulate immunoglobulin IgE production.

Epitope See Antigenic determinant.

Etiology The study of the origin and progression of a disease.

Eukaryote A cell or organism with a distinct membrane-bound nucleus as well as specialized membrane-based organelles (see also Prokaryote).

Ex vivo An experiment performed on living tissue extracted from an organism that is to be subsequently reintroduced to that organism (e.g., removing stem cells from human bone marrow prior to gene therapy and their subsequent reintroduction to the bloodstream).

Exon The region of DNA within a gene that codes for a polypeptide chain or domain. Typically a mature protein is composed of several domains coded by different exons within a single gene.

Exonuclease An enzyme that cleaves DNA at the free 5′ or 3′ ends by breaking the phosphodiester bonds of the backbone.

Fab The fragment of an antibody containing the antigen-binding site, generated by cleavage of the antibody with the protease papain, which cuts at the "hinge" region of the Y-shaped antibody molecule and produces two Fab fragments.

Fc The fragment of the antibody produced after papain digestion that does not contain the antigen-binding site (the stem of the "Y"). The Fc fragment contains the C-terminal domains of the two heavy chains.

Fetal antigen Cell surface antigens expressed on embryonic tissues that is re-expressed during tumor growth.

Fluoroscein A yellow dye used to label antibodies for fluorescence studies.

Fluorescence The phenomenon of delayed emission of light of a specific wavelength by some molecules after absorption of light of a different wavelength. Fluorescent molecules are used in numerous assays as conjugates with antibodies or other ligands.

Fluorescence *in situ* hybridization (FISH) A variation of the DNA/RNA hybridization procedure in which the denatured DNA is kept in place in the chromosome and is then challenged with RNA or DNA extracted from another source to which a fluorescent tag has been attached. The advantage of maintaining the DNA in the chromosome is that the specific chromosome (or chromosomes) containing that region of DNA is established by detecting the fluorescence.

Force field Field describing the forces between a ligand and its receptor.

Frameshift A deletion, substitution, or duplication of one or more bases that causes the reading-frame of a structural gene to go out of phase.

G proteins Set of related proteins involved in signal transduction.

Gamma globulins Serum proteins with "gamma" electrophoretic mobility that interact with the immunglobulins.

Gel electrophoresis A technique by which molecules are separated by shape, size, or charge by passing them through a gel under the influence of an external electric field.

Gene Classically, a unit of inheritance. In practice, a gene is a segment of DNA on a chromosome that has a specific effect on an organism that is directly attributable to it. Genes may be structural (code for the structure of a polypeptide, protein, or RNA) or regulatory (code for a promoter or enhancer region).

Gene library A collection of cloned DNA fragments created by restriction endonuclease digestion that represent part or all of an organism's genome.

Genetic code The mapping of DNA or RNA triplets (3 bases), usually known as codons, into the 20 amino acids as well as the start and stop codons.

Genetic marker Any gene that can be readily recognized by its phenotypic effect and that can be used as a marker for a cell, chromosome, or individual carrying that gene. Also any detectable RFLP used to identify a specific gene linked to it.

Genome The complete genetic content of an organism.

Genotype Strictly, all of the genes possessed by an individual. In practice, the particular alleles present in a specific genetic locus.

Germ line Those cells that are specifically involved in reproduction; in particular, their genes after the reduction in chromosome number (cf. Somatic cells).

Glycosylation The addition of carbohydrate groups, such as to polypeptide chains.

Golgi body (Golgi network) A cellular organelle formed by invaginations of the endoplasmic reticulum to form a series of closed sacs. Golgi bodies are sites of protein assembly and post-translational modification.

Graft rejection A T-cell-mediated immune reaction elicited by grafting genetically dissimilar tissues to a recipient. The result of the reaction is typically the rejection of the transplanted tissue.

Granulocyte White blood cells of the granulocytic leukocyte series including the esoinophils, neutrophils, and basophils.

Growth factor A class of small proteins that signal growth and division to certain cell types by binding to a cell surface receptor and signaling the expression of genes involved in cell division.

Guanine (G) One of the nitrogenous purine bases found in DNA and RNA.

Gyrase An enzyme, one of the family known as topoisomerases, that can alter the topological state of supercoiled DNA through its nicking and ligating activities. Gyrase can induce negative supercoils into closed circular DNA.

H-2 The mouse major histocompatibility complex (MHC) located on chromosome 17.

Hairpin A double-helical region in a single DNA or RNA strand formed by the hydrogen bonding between adjacent inverse complementary sequences along the nucleic acid strand.

Haploid A cell or organism containing only one complete set of chromosomes (cf. Diploid).

Hapten A compound of low molecular weight (<700 dal) that becomes immunogenic only when complexed with a carrier protein or cell.

Helicase An enzyme that unwinds the DNA double helix during the time of DNA replication.

Helper T cells A class of T cells that triggers B cells to produce antibodies and also helps generate cytotoxic T cells. These cells are identified by expression of the CD4 protein marker on their cell surface.

Hemagglutination The agglutination of red blood cells under the effect of integrins.

Histamine A vasoactive amine released from granules within mast cells and basophils that causes vasodilation and smooth muscle contraction. One of the major mediators in immediate hypersensitivity or allergic reactions.

Histocompatibility Literally, the ability of different tissues to "get along"; strictly, identity in all transplanation antigens. A requirement for the prevention of graft or organ rejection.

Histones Basic proteins found in eukaryotes that package and regulate the DNA in chomosomes.

HLA complex Another name for the MHC in humans, refers to the "human leukocyte antigen" complex located on chromosome 6.

Homeobox A highly conserved region in a homeotic gene composed of 180 bases (60 amino acids) that specifies a protein domain (the homeodomain) that seems to serve as a master genetic regulatory element.

Homeodomain A 60 amino-acid protein domain coded for by the homeobox region of a homeotic gene.

Homeotic gene A gene that controls the activity of other genes involved in the development of a body plan. Homeotic genes have been found in organisms ranging from plants to humans.

Homologous Strictly, corresponding genetic loci. In common usage, two or more genes or alleles sharing a significant degree of similarity in their DNA sequence.

Human anti-murine antibody response (HAMA) An immune response generated in humans to therapeutic antibodies raised in murine cells, such as mouse or rat.

Humanization The modification of a murine molecule or cell to make it appear "human" to the human immune system. Commonly used to describe the process whereby antibodies generated in mice or rats have sufficient homologous amino acids substituted by their human counterparts so as to prevent an immunogenic response.

Human immunodeficiency virus (HIV) The general name for a family of retroviruses that cause acquired immunodeficiency syndrome (AIDS).

Hybridoma A hybrid cell resulting from the fusion of an antibody-secreting B cell with an immortal myeloma (malignant cancer) cell; hybridoma lines can be cultured *in vitro* and will secrete a requisite antibody without stimulation while proliferating continuously.

Hydrogen bond A weak chemical interaction between an electronegative atom (e.g., nitrogen or oxygen) and a hydrogen atom that is covalently attached to another atom. The bond that maintains the two helixes of DNA together and that is the primary interaction between water molecules.

Hydrophilicity Literally, water-loving; the degree to which a molecule is soluble in water. Hydrophilicity depends to a large degree on the polarizability of the molecule and its ability to form transient hydrogen bonds with polar water molecules.

Hydrophobicity Literally, water-hating; the degree to which a molecule is insoluble in water and hence is soluble in lipids. If a molecule lacking polar groups is placed in water, it will be entropically driven to finding a hydrophobic environment such as the interior of a protein or a membrane.

Idiotope Antigenic determinant of an idiotype (see Idiotype).

Idiotype Antibody variants localized to the variable portion of an immunoglobulin that are recognized by their antigenic determinants. The determinants are composed from the antigen-combining site or CDRs. Every unique antigenic determinant has a specific antibody with its own unique idiotype.

Immune tolerance The ability of the immune system to attack foreign antigens yet remain inactive against self-proteins; this characteristic can apply to both T cells and B cells.

Immunogen A substance that induces an active immune response. Note that while all immunogens are antigens, not all antigens are immunogenic.

Immunoglobulin A member of the globulin protein family consisting of two light and two heavy chains linked by disulfide bonds. All antibodies are immunoglobulins.

Immunosuppression (Immune suppression) The downregulation of the immune system by regulatory systems or external agents such as drugs or radiation.

***In situ* hybridization** A variation of the DNA/RNA hybridization procedure in which the denatured DNA is in place in the cell and is then challenged with RNA and DNA extracted from another source. (See also Fluorescence *in situ* hybridization.)

In vitro Occurring outside the living organism (literally, in glass). Typically an experiment performed in a test tube or other artificially designed environment (cf. *In vivo*).

In vivo Occurring within a living organism (cf. *In vitro*).

Inflammation Response resulting from tissue injury that involves the increase in blood supply to the site of the injury, increased capillary permeability, and movement of leukocytes from capillaries into the site.

Interphase The condition of the nucleus of a cell in between nuclear divisions (the "resting" phase). A nucleus that has not yet entered either the mitotic or meiotic cycle.

Integrins Family of cell surface receptors that are transmembrane glycoproteins consisting of a small C-terminal intracellular domain and a large extracellular domain. The extracellular domain interacts with a variety of ligands, resulting in cell–cell and cell–matrix adhesion during hemostatis, wound healing, and immune defense mechanisms.

Interleukins Family of proteins secreted by leukocytes allowing them to communicate. Members of the cytokines family.

Introns Nucleotide sequences found in the structural genes of eukaryotes that are noncoding and interrupt the sequences containing information that codes for polypeptide chains. Intron sequences are spliced out of their RNA transcripts before maturation and protein synthesis (cf. Exons).

Inverted repeats Two identical DNA sequences oriented in opposite directions on the same molecule. Adjacent inverted repeats are known as a palindrome.

Isotypes Different classes of protein molecules that otherwise share the same structure and size. Antibodies exist in five different isotypes, IgG, IgM, IgD, IgE, and IgA.

Isozymes Two or more enzymes capable of catalyzing the same reaction but differing slightly in their specificity due to differences in their structures and consequently their efficiencies under different environmental conditions.

Joining Linking of exons or DNA segments in somatic cell genes that are separated by introns in eukaryotes.

Kappa chains A family of antibody light chains characterized by their constant-domain amino-acid sequence.

Karyotype The constitution of chromosomes (typically number and size) in a cell or individual.

Kinins Small molecules released during an anaphylactic reaction that induce dilation of blood vessels and contraction of smooth muscle.

Lambda chains Family of antibody light chains characterized by their particular constant-domain amino-acid sequence (cf. Kappa chains).

Lagging strand The DNA strand that is assembled discontinuously during replication in the direction away from the replication fork.

Leader sequence (signal sequence) A short sequence at the N-terminal end of a protein whose function is to translocate that protein across a membrane such as the endoplasmic reticulum of cell membrane. After translocation the leader sequence is cleaved from the remainder of the protein.

Leading strand The DNA strand that is assembled continuously during replication in the direction toward the replication fork.

Least squares Statistical method of correlation that minimizes the square of the differences between two variables.

Ligand Any small molecule that binds to a protein or receptor; the cognate partner of a cellular protein, enzyme, or receptor.

Linkage The association of genes or genetic loci on the same chromosome. Genes that are linked together tend to be transmitted together.

Linkage map A genetic map of a chromosome or genome delineated by mapping the positions of genes by their linkage to readily identifiable genetic loci.

Linker (DNA) A short piece of DNA with a restriction endonucleotide recognition sequence built into its sequence that is used to convert, for example, a blunt-end DNA fragment into a sticky-end fragment.

Lipofection The procedure by which foreign DNA is introduced into mammalian cells via liposomes (see Liposomes).

Liposomes Spherical droplets (microvesicles) composed of a lipid bilayer shell into which therapeutic agents may be encapsulated. Liposomes fuse with cellular membranes of target cells when injected, enabling their encapsulated molecules to be liberated into the cellular milieu.

LMP Low molecular weight proteins that are part of the proteasome complex.

Locus The specific position occupied by a gene on a chromosome. At a given locus any one of the variant forms of a gene may be present.

Lod (log-of-odds) Score—measure of the extent of linkage between two genetic markers.

Long-terminal repeats (LTRs) Sequences that are several hundred bases long that are directly repeated at the ends of the DNA of many retroviral genomes.

Lymph nodes Small pea-sized organs distributed about the body composed of lymphoid and accessory cells that act as filters for the lymphoid system. The site of localized immune response.

Lymphocytes Spherical cells of the lymph system associated with specific immunity.

Lymphoid tissue Body tissue composed primarily of lymphocytes. Lymph tissue includes the spleen, lymph nodes, thymus, tonsils, adenoids, and circulating lymph fluid.

Lymphoma Cancer of the lymphocytes or lymph tissue.

Lysis The bursting of a cell following disruption or dissolution of the cell membrane.

Lysogenic virus A virus that is dormant in its host cell during which its genome is stably integrated into the genome of the host cell.

Lysosome A cytoplasmic organelle containing hydrolytic enzymes. The enzymes remain inactive until release during intracellular digestion.

Lysozyme An enzyme found in tears, saliva, and nasal secretions that lyses mainly gram positive bacteria. Lysozyme cleaves the muramic acid beta(1-4)-*N*-acetylglucosamine linkage found in the cell walls of these bacteria.

Lytic virus A virus that causes lysis of the host cell.

Macrophage A diverse group of immune cells that engulf and destroy foreign particles. Macrophages play a role in both nonspecific immunity (removal of foreign material) and specific immunity (processing and presenting antigen to T cells).

Malignant transformation The conversion of a noncancerous cell to the cancerous state by some trigger such as the acquisition of an oncogene.

Malignant tumors Tumors that have the capability of invading the surrounding nontumorous tissues.

Major histocompatibility complex (MHC) A region of the genome (on the short arm of chromosome 6 in humans) containing a number of genes that control the processing and presentation of antigens and other genes that play an important role in the immune system. The presenting gene products are the class I and class II MHC molecules that also play an important role in tissue graft rejection.

Map unit A measure of genetic distance between two linked genes that corresponds to a recombination frequency of 1%.

Mast cells The tissue-bound equivalent of circulatory basophils (see basophils).

Meiosis A process within the cell nucleus that results in the reduction of the chromosome number from diploid to haploid through two divisions in germ cells prior to fusion or sexual reproduction.

Melting The denaturation of double-stranded DNA into two single strands by the application of heat. (Denaturation breaks the hydrogen bonds holding the double-stranded DNA together.)

Memory In the immune system, a phenomenon whereby immune system responds faster and more powerfully to an antigen on exposures subsequent to the primary exposure. Also a heightened secondary immune response.

Mendelian trait (Mendelian inheritance) A trait that is transmitted in accordance with Mendel's laws.

Messenger RNA (mRNA) The complementary RNA copy of DNA formed on a single-stranded DNA template during transcription that is processed into a sequence carrying the information to code for a polypeptide domain.

Metaphase The stage during nuclear division when the chromosomes are arranged at the equatorial plane of the spindle.

Mimetics Compounds that mimic the function of other molecules due to their high degree of structural (conformational) similarity.

Mismatch repair The enzyme-based correction of nucleotide mispairing during DNA replication that involves removal of defective single-stranded segments (containing the mismatch) followed by synthesis of new segments. An entire mismatch repair machinery exists in the cell to maintain the fidelity of DNA replication.

Missense mutation A point mutation in which one codon (triplet of bases) is changed into another designating a different amino acid.

Mitochondrion A cytoplasmic organelle containing an invaginated double membrane to which are attached a number of the enzymes involved in aerobic respiration in eukaryotes.

Mitogen Any agent that induces mitosis in somatic cells.

Mitosis The nuclear division that results in the replication of the genetic material and its accurate redistribution into each of the daughter cells during cell division.

Monocyte The form of macrophages that circulate in the bloodstream.

Monoclonal antibodies Identical antibodies produced from a single line (clone) of B cells. The resultant molecules are identical in sequence and hence affinity, binding specificity, idiotype, etc. Monoclonal antibodies were originally produced by fusing a chosen B-cell line with an immortal myeloma cell line to produce so-called hybridomas, immortal cells that secrete only the single antibody type of the selected B-cell clone.

Monomer A single unit of any biological molecule or macromolecule, such as an amino acid, nucleic acid, polypeptide domain, or protein.

Multigene family A set of genes derived by duplication of an ancestral gene followed by independent mutations resulting in a series of independent genes either clustered together on a chromosome or dispersed throughout the genome.

Mutagen Any agent that can cause an increase in the rate of mutations in an organism.

Mutation An inheritable alteration to the genome that includes genetic (point- or single-base) alterations or larger-scale alterations such as chromosomal deletions or rearrangements.

Naked DNA The term used to describe pure, isolated DNA devoid of any proteins that may bind to it.

Natural selection The differential reproduction of alleles occurring from one population to the next over differing generations that results in the increased presence of some alleles and the decreased presence of others.

Natural killer (NK) cells A group of Fc receptor-bearing cytotoxic T cells having the inherent ability to kill virally infected cells and some tumor cells without prior immunization. NK cells can also destroy targets coated with IgG antibodies.

Negative control Genetic regulation whereby a regulatory element represses transcription of one or more specific structural genes.

Neoplasm Any new or abnormal cellular growth, such as a tumor or a cancer.

Nonobese diabetic (NOD) mouse An inbred mouse strain that spontaneously develops a disease very similar to insulin-dependent diabetes mellitus.

Nonsense codon A codon that signals the termination of the polypeptide chain.

Nonsense mutation A point mutation in which a codon specific for an amino acid is converted into a nonsense codon.

Northern blotting A technique to identify RNA molecules by hybridization that is analogous to Southern blotting (see Southern blotting).

Nuclease Any enzyme that can cleave the phosphodiester bonds of nucleic acid backbones.

Nucleolus A body within the nucleus of the cell containing genes for ribosomal RNA and their associated processing proteins.

Nucleoside A five-carbon sugar covalently attached to a nitrogen base.

Nucleosome The basic structural unit of chromatin consisting of about 200 base pairs of DNA wrapped around a complex of 8 histone proteins.

Nucleotide A nucleic acid unit composed of a five-carbon sugar joined to a phosphate group and a nitrogen base.

Okazaki fragments Short DNA chains resulting from discontinuous replication from one of the template strands of double-stranded DNA during replication.

Oligonucleotide A short molecule consisting of several linked nuceotides (typically between 10 and 60) covalently attached by phosphodiester bonds.

Oncogene A gene that can initiate and maintain a tumerous state in an organism. Oncogenes are derived from protooncogenes and are present in most cells.

Oncogenesis The development of a neoplasm or malignancy.

Open road frame (ORF) Contiguos stretch of amino acids present in a DNA sequence that potentially codes for a protein.

Operator A segment of DNA that interacts with the products of regulatory genes and facilitates the transcription of one or more structural genes.

Operon A unit of transcription consisting of one or more structural genes, an operator, and a promoter.

Opsonin A substance capable of enhancing phagocytosis. Opsonins include antibodies and the C3b molecule of the complement system.

Oral tolerance Term used to describe the phenomenon of tolerance to a foreign protein obtained by ingestion of that protein. The mechanisms of this effect remain to be fully elucidated.

Organelle A subcellular component usually containing particular enzymes involved in a specialized role.

Palindrome A region of DNA with a symmetrical arrangement of bases occurring about a single point such that the base sequences on either side of that point are identical if the strands are both read in the same direction.

Passive immunity Protection (acquired immunity) due to antibodies or T cells acquired from another immune individual.

Pathogen Any foreign agent or organism that is harmful to the body.

Peptide A short stretch (usually less than 50) of amino acids covalently coupled by a peptide (amide) bond.

Peptide bond (amide bond) A covalent bond formed between two amino acids when the amino group of one is linked to the carboxy group of another resulting in the elimination of one water molecule. Peptide bonds are planar.

Peptidyl transferase An enzyme found in the large ribosomal subunit that catalyzes the formation of peptide bonds between two adjacent amino acids.

Peyer's patches Nodules of lymphoid tissues located in the submucosa of the small intestine. They typically contain B cells, plasma cells, and germinal centers.

Phage A virus that infects bacterial cells.

Phage display A technique in which phage are engineered to fuse a foreign peptide or protein with their capsid (surface) proteins and hence display it on their cell surfaces. The immobilized phage may then be used as a screen to see what ligands bind to the expressed fusion protein.

Phagocytosis The engulfment of material into phagosomes by phagocytic cells such as macrophages.

Pharmacophore The three-dimensional spatial arrangement of atoms, substituents, functional groups, or chemical features that together are sufficient to describe the pharmcaolgically active components of a molecule or molecule series.

Phenotype Any detectable feature of an organism that is the result of one or more genes.

Phospholipase A3 A membrane-associated enzyme that releases arachidonic acid from membrane lipid, the first step in the cycloxygenase and lipoxygenase pathways, leading to the formation of prostaglandins and leukotrienes, respectively.

Pinocytosis The engulfment of liquids or very small particles by phagocytic cells.

Plasmid Any replicating DNA element that can exist in the cell independently of the chromosomes. It is most commonly found in bacterial cells.

Pleitropy The multiple effects on an organism's phenotype due to a single gene or allele.

Point mutation A mutation in which a single nucleotide in a DNA sequence is substituted by another.

Poly(A) tail The stretch of Adenine (A) residues at the 3′ end of eukaryotic mRNA that is added to the pre-mRNA as it is processed, before its transport from the nucleus to the cytoplasm and translation at the ribosome.

Polyethylene glycol (PEG) An immunologically inert substance used to coat molecules or fuse cells.

Polygenic inheritance Inheritance involving alleles at many genetic loci that may interact with environmental factors.

Polymorphism The existence of a gene in a population in at least two different forms at a frequency far higher than that attributable to recurrent mutation alone.

Polypeptide A single chain of covalently attached amino acids joined by peptide bonds. Polypeptide chains usually fold into a compact, stable form (a domain) that is part or all of the final protein.

Polysome An assembly of ribosomes actively translating the same mRNA strand into polypeptide chains.

Polytene chromosome A chromosome packaged as a group of many associated strands of chromatids becoming very large and easy to discern by light microscopy. It is found in the salivary glands of fruit flies.

Population An interbreeding group of organisms of the same species.

Positive control Genetic regulation whereby the product of a regulatory gene is required to activate transcription of specific structural genes.

Post-transcriptional modification Any alteration made to pre-mRNA before it leaves the nucleus and becomes mature mRNA.

Primer A short oligonucleotide that provides a free 3′ hydroxyl for DNA or RNA synthesis by the appropriate polymerase (DNA polymerase or RNA polymerase).

Probe Any biochemical that is labeled or tagged in some way so that it can be used to identify or isolate a gene, RNA, or protein.

Prokaryote An organism or cell that lacks a membrane-bounded nucleus. Bacteria and blue-green algae are the only surviving prokaryotes (cf. Eukaryote).

Promoter A segment of DNA that contains the start signals for RNA polymerase and hence promotes transcription at the start of a structural gene. The binding site of transcription factors that regulates gene expression.

Prostaglandins A variety of pharmacologically active derivatives of arachidonic acid that causes increased vascular permeability, smooth muscle contraction, and bronchial constriction and is an important mediator of hypersensitivity.

Proteasome A large, multisubunit, multicatalytic complex proposed to play a role in antigen processing.

Proto-oncogene A cellular gene that can be converted to an oncogene as the result of somatic mutation or recombination with a viral gene.

Provirus The DNA form of a virus that can integrate into the chromosome of its host cell.

Pseudogene A gene that closely resembles a known functional gene at another locus yet has become nonfunctional due to an accumulation of defects that prevent normal transcription or translation.

Purine A nitrogen-containing compound with a double-ring structure. The parent compound of Adenine and Guanine.

Pyrimidine A nitrogen-containing compound with a single six-membered ring structure. It is the parent compound of Thymidine and Cytosine.

Radioimmunoassay (RIA) A competitive immunological procedure for measuring very low concentrations of antigens or antibodies by using radioactively labeled antigens as competitors.

Reading frame A sequence of codons beginning with an initiation codon and ending with a termination codon, typically of at least 150 bases (50 amino acids) coding for a polypeptide or protein chain.

Reanneal To form duplex DNA or RNA molecules from complementary single-stranded chains.

Receptor The binding site of a signaling molecule, ligand, or chemical. Typically receptors are proteins and are usually bound to a membrane.

Recessive Any trait that is expressed phenotypically only when present on both alleles of a gene.

Recombinant DNA DNA molecules resulting from the fusion of DNA from different sources. Also the technology employed for splicing DNA from different sources and for amplifying heterogenous DNA.

Recombination A new combination of alleles resulting from the rearrangement occurring by crossing over or by independent assortment (see Crossing over).

Regulatory gene A DNA sequence that functions to control the expression of other structural genes by producing a protein that modulates the synthesis of their products.

Renaturation The reconstitution of a molecule (typically DNA or protein) from its denatured state by the reassociation of its constituent chains.

Repetitive DNA Repeated DNA sequences that may occur in hundreds or even thousands of copies in the chromosomes of eukaryotes.

Replication The synthesis of a macromolecule (e.g., DNA) from a template molecule.

Replication fork The Y-shaped region within a double-stranded DNA helix undergoing replication where local denaturation and synthesis occurs.

Replicon A unit of DNA replication with a specific point of origin.

Repressor The protein product of a regulatory gene that combines with a specific operator (regulatory DNA sequence) and hence block the transcription of structural genes in an operon.

Restriction enzyme (restriction endonuclease) A type of enzyme that recognizes specific DNA sequences and produces cuts on both strands of DNA containing those sequences.

Restriction fragment length polymorphisms (RFLPs) Variation within the DNA sequences of organisms of a given species that can be identified by fragmenting the sequences using restriction enzymes.

Restriction map A physical map or depiction of a gene (or genome) derived by ordering overlapping restriction fragments produced by digestion of the DNA with a number of restriction enzymes.

Retrovirus A virus that has RNA as its genetic material and uses reverse transcriptase to make a DNA copy prior to its integration into the hosts genome.

Reversion A change to a mutant allele that reverts it back to the wild-type allele.

Reverse transcriptase A DNA polymerase that can synthesize a complementary DNA (cDNA) strand using RNA as a template (a so-called RNA-dependent DNA polymerase).

Reverse turn Turn formed by four amino acids that allows adjacent strands of beta sheet to connect.

Rheumatoid arthritis (RA) A type III hypersensitivity due to the deposition of immune complexes at the joints leading to joint inflammation and damage, thus an autoimmune disease.

rho factor A protein required to halt the transcription of genes.

Ribonuclease (RNase) An enzyme that hydrolyzes (breaks down) RNA.

Ribosomal RNA (rRNA) Class of RNA molecules that play a structural and functional role in protein synthesis and is found in the small and large subunits of the ribosome.

Ribosome A complex cellular organelle composed of RNA and protein that serves as the site for protein synthesis from a mature mRNA transcript previously created in the nucleus.

Ribonucleic acid (RNA) A category of polynucleotides in which the component sugar is ribose and consists of the four nucleotides Cytosine, Uracil, Guanine, and Adenine.

Ribozymes Catalytic or autocatalytic RNA molecules that have arisen through either natural or artificial molecular evolution. Ribozymes that catalyze the making and breaking of phosphodiester bonds are speculated as being the earliest molecules of life.

RNA polymerase An enzyme that catalyzes the synthesis of RNA from ribonucleoside triphosphate precursors from a template DNA strand.

Rolling circle A model for the manner of replication of certain DNA molecules for circular DNA.

S phase Portion of the cell cycle during which DNA replication (synthesis) takes place.

Sarcoma A malignant tumor of mesodermal origin, such as in the connective tissue, bone, or muscle.

Satellite Terminal end of a chromosome arm produced by a constriction that is connected to the rest of the chromosome by only a very small region.

Satellite DNA DNA from a eukaryotic cell with a different sedimentation constant from the bulk of the nuclear DNA, and hence it equilibrates at a different position ("satellite bands") from the main band of DNA following density gradient centrifugation.

Secondary response Immune response following second exposure to a given antigen. In general, secondary responses are quicker, more protective, and last longer than primary responses.

Selective pressure The operation of natural selection on the allele frequency in a population resulting in the increase in frequency of favored alleles and the decrease in frequency of unfavored alleles.

Semiconservative replication The method of replication of double-stranded DNA that ensures that one parent strand of DNA ends up in each daughter cell, together with one newly synthesized strand formed using it as a template.

Sense strand The strand of double-stranded DNA that acts as the template strand for RNA synthesis. Typically only one gene product is produced per gene, reading from the sense strand only. (Some viruses have open-reading frames in both the sense and the antisense strands.)

Serine protease A family of protein-cutting proteins, each with a serine residue at the active site. The enzymes, although sharing little sequence homology, have a distinctive catalytic arrangement of residues at the active site that allow them to cleave peptide (amide) bonds.

Sex chromosomes The chromosomes involved in sex determination (e.g., X and Y in humans) that show a difference in number and morphology between the sexes.

Serum Liquid part of coagulated blood (after conversion of fibrinogen to fibrin) after removal of cells and fibrin.

Serology Study of *in vitro* interactions of antibodies with antigens.

Serotonin Molecule that has a variety of functions: It is a neurotransmitter, it is present on the mast cell and in platelets, and it contributes to hypersensitivity reactions.

Severe combined immunodeficiency (SCID) A group of diseases in which the patient exhibits a low level of leukocytes (leukopenia) and an impaired or nonexistent immune response. Typically this leads to death by infection.

Shotgun cloning The cloning of an entire gene segment or genome by generating a random set of fragments using restriction endonucleases to create a gene library that can be subsequently mapped and sequenced to reconstruct the entire genome.

Sigma subunit One domain of the multidomain enzyme RNA polymerase that is essential for the recognition of signals in the DNA specifying the start of transcription.

Signal sequence (leader sequence) A short sequence added to the amino-terminal end of a polypeptide chain that forms an amphipathic helix, allowing the nascent polypeptide to migrate through membranes such as the endoplasmic reticulum or the cell membrane. It is cleaved from the polypeptide after the protein has crossed the membrane.

Signal transduction The pathway whereby stimuli external to the cell are transmitted from the cell surface to the cell nucleus. Signal transduction pathways usually rely on a cell surface receptor that reacts to a particular external "message"; the receptor then triggers a "second messenger" within the cell that activates a cascade of protein–protein interactions resulting in the selective transcription of genes within the nucleus.

Silent mutation A mutation in which a codon for a specific amino acid is altered into one that designates the same amino acid.

Small ribonuclear particle (snRNP) A complex found in the nucleus composed of one RNA molecule and several proteins that plays a role in the processing of pre-mRNA.

Somatic cells All the cells of the body except those involved in reproduction (the germ line cells).

Southern blotting A procedure for the identification of DNA by transmitting a fragment isolated on an agarose gel to a nitrocellulose filter where it can be hybridized with a complementary "probe" sequence.

Spindle An aggregation of microtubules essential for the positioning and movement of the chromosomes during nuclear division.

Spliceosome A large complex of small ribonuclear proteins (snRNPs) and other proteins that together process pre-mRNA into a mature mRNA transcript.

Splicing The joining together of separate component parts. For example, RNA splicing in eukaryotes involves the removal of introns and the stitching together of the exons from the pre-mRNA transcript before maturation.

Stem cells Immature (precursor) cells from which mature cells arise, typically the cells of the bone marrow that are the progenitors of the T cells and B cells of the immune system.

Sticky-end ligation The attachment of DNA to a DNA fragment whose ends contain "overhangs" in their individual DNA strands and hence templates for the hybridization of fragments with the complementary base sequence. Sticky-ends are typically made by restriction enzymes that cleave DNA assymetrically.

Substituent A chemical group that can be bound at a particular site of a chemical scaffold or congeneric (homologous) series.

Superantigen A protein that is capable of stimulating a large population of T cells by bridging the outside of the MHC and a particular T-cell receptor beta chain.

Suppressor T cells A subset of T lymphocytes responsible for the down regulation of antibody production by B cells and cell-mediated reactions by other T cells.

Syncytia Plaques or clumps of cells (usually dead) that have been infected with a foreign agent or virus such as HIV.

Systemic lupus erythromatosus (SLE) An autoimmune disease involving antibodies to nuclear and other intracellular components of DNA-related machinery.

T-cell receptor (TCR) A disulfide-bonded heterodimer that recognizes a foreign peptide and MHC molecule simultaneously.

T lymphocyte (T cell) A white blood cell, matured at the thymus, that is able to recognize foreign antigens presented by MHC molecules on cell surfaces and stimulate an immune response bringing about their elimination from the body.

Tautomerism The shifting of the location of a proton from one region of a molecule to another that in turn alters its chemical properties.

Telomere A region of the chromosome marking the extreme end of each chromosome arm.

Template A macromolecule that serves as the blueprint or mould for the synthesis of another molecule.

3′end The end of a nucleic acid chain containing a free 3′-OH (hydroxyl) group.

Thymic education The process by which T cells developing in the thymus are screened for potentially harmful self-reactive T cells which are ignored while useful T cells are promoted.

Thymine A pyrimidine base found in DNA but not in RNA.

Thymopoientin A hormone produced by epithelial cells of the thymus that regulates early T-cell differentiation and proliferation.

Thymus A primary lymphoid organ that serves as the site for T-cell development.

Tolerance In the immune system, specific lack of responsiveness to selected antigens.

Topoisomerase Enzymes that convert topologically constrained DNA from one topological form to another.

Transcript The single-stranded mRNA chain that is assembled from a DNA template.

Transcription The assembly of complementary single-stranded RNA on a DNA template.

Transcription factors A varied group of regulatory proteins that are required for transcription in eukaryotes.

Transduction The transfer of genetic material from one cell to another by a viral vector.

Transfection The uptake in culture by recipient cells of exogenous DNA. The cells are either treated with calcium phosphate or subjected to an electric field (electroporation) to make their membranes more permeable. Standard procedure for introducing foreign DNA into mammalian cells.

Transfer RNA (tRNA) A small RNA molecule that recognizes a specific amino acid, transports it to a specific codon in the mRNA, and positions it properly in the nascent polypeptide chain.

Transformation A genetic alteration to a cell as a result of the incorporation of DNA from a genetically different cell or virus; transformation can also refer to the incorporation of extraneous DNA by bacterial cells during genetic manipulation.

Transition A point mutation where a purine is replaced by another purine, or a pyrimidine with another pyrimidine.

Translation The process of converting RNA to protein by the assembly of a polypeptide chain from an mRNA molecule at the ribosome.

Translocation A chromosomal alteration in which a chromosome segment or arm is transposed to a new location.

Transposon A genetic unit capable of moving from one chromosome site to another or from one replicon to another.

Transversion A point mutation in which a purine is replaced by a pyrimidine, and vice versa.

Upstream Region of DNA template toward the 3′ end from a site of interaction, such as the initiation of transcription.

Uracil Nitrogenous pyrimidine base found in RNA but not DNA.

Vaccination The inoculation of an animal with an attenuated antigen in order to generate a protective immune response against the organism bearing that antigen.

Vector Any agent that transfers material (typically DNA) from one host to another. Typically DNA vectors are autonomous DNA elements (e.g., plasmids) that can be manipulated and integrated into a host's DNA.

Virion Complete virus particle including nucleic acid core and protein (capsid) coat prior to budding from its host cell.

Wild type Form of a gene or allele that is considered the "standard" or most common type found in nature.

Wobble hypothesis The concept that the base at the 5′ end of the tRNA anticodon is free to move spatially and can thus form hydrogen bonds with more than just one kind of base at the 3′ end of a codon in the mRNA.

X chromosome In mammals, the sex chromosome that is found in two copies in the homogametic sex and one copy in the heterogametic sex.

Z-DNA A form of DNA existing as a left-handed double helix (the phosphate-sugar backbone forms a left-handed zigzag course) which may play a role in gene regulation.

INDEX

Accessory molecules, targeting, 178–179
Angiotensinogen converting enzyme inhibitor series, QSAR analysis, 131–132
Adenoviral vectors:
  gene insertion, 45, 49
  generation, 49
Alzheimer's disease, 29–31
Antibody:
  engineering, immunotherapy, 114, 116–117
  fragments, 117–118
  genetically engineered, cancer diagnostics and therapy, 119–120
  idiotypic, 118–119
  rational drug design, 118–119
  therapeutic and diagnostic, 117–118
Antibody genes, immune system, 157
Antigen, oral or nasal administration, 172
Antigen-building determinants, 114–117
Anti-P185her2 antibody, 114–117
Anti-proliferative activity, 114–117
Autoantigens, insulin-dependent diabetes mellitus, 169–172
Autosomal dominant diseases, 7, 10

Binary masking, 92–94
Bioinformatics, role in genome-based therapy, 20–24
Biopharmaceutical companies, 186–190
Biotechnology companies, drug development research, 74–75
Bovine pancreatic trypsin inhibitor, 88–89

Cancer, diagnostics and therapy, genetically engineering antibodies, 119–120
Carriers, 8
Cationic liposomes, gene insertion, 50–51
CD4:
  mimetic to, 123
  T cells, 158
    interactions, 164–165
Cell adhesion molecules, targeting, 179
Central tolerance, 166
Cloning:
  disease genes, 8–13
    EST sequencing, 9, 11–12
    Human Gene Map, 12–13
  positional, 6
Clustering methods, 104–105
Co-inheritance, 6
Combinatorial chemistry:
  drug development, 72–77
  library, compound identification, 81, 83
  principles, 80–83
Comparative molecular field analysis (CoMFA), 131, 135, 137
Computational chemistry:
  quantitative structure-activity relationships, 129, 133, 135–137
  role in drug design, 109–110
Conformational searching, drug design, 128
Cre/Lox system, 61–62
Cyclosporine, 177–178
Cystic fibrosis, 7
  molecular diagnostics and therapeutics, 25–27
  occurrence of carriers, 37
  position cloning, 6
Cytosolic antigens, processing, 164
Cytokines, 160
  inflammatory, 180
  Th1-Th2 balance, 168–169

D1 agonists, potent and selective, design, 130–131
Databases, 3D, computational screening, 137–138
Delivery systems, 39–54
  gene insertion, cellular modalities, 39–41
  nonviral systems, 50–54
De novo drug design, based on target 3D structures, 141

Diabetes mellitus, insulin-dependent, 169–172
methods used to prevent, 182
Differential display technique, 15–17
Directed evolution, proteins of therapeutic significance, 88–89
Disease genes:
analysis, 24–31
Alzheimer's disease, 29–31
cystic fibrosis, 25–27
Huntington's disease, 24–25
retinosis pigmentosa, 27–29
methods for tracking, 18–19
Disease inheritance, modes, 8, 10
Diversity metrics, 104
"Diversomer" synthesis, organic diversity, 102–103
Drug design, 109–148
computational chemistry role, 109–110
computational screening, 3D databases, 137–138
pharmacophore developed from series of delta-selective opioid cyclic peptides, 130
potent and selective D1 agonists, 130–131
protein and antibody engineering, 110–120
rational, antibodies, 118–119
small-molecule, computational chemistry, 124–147
classical molecular mechanics, 126–128
computational screening of 3D databases, 137–138
conformational searching and pharmacophore mapping, 128–129
de novo drug design, 141
quantitative structure-activity relationships, 129, 133–137
quantum mechanics, 126
structure-based drug design, 138–141
without target 3D structures, 124–126
structure-based, xviii, 142–147
HIV protease inhibitors, 145–147
potent antagonists to human growth hormone receptor, 142–143
purine nucleoside phosphorylase inhibitors, 143
thymidilate synthase inhibitors, 143–145
Drug development, combinatorial chemistry, 72–77
Drug discovery, 68–106
cycle, 110–111
directed molecular evolution, 76–77
molecular diversity, *see* Molecular diversity
therapeutic targets, 68–72
extracellular, 71–72
intracellular, 69, 71
Duchenne's muscular dystrophy, 9

Encephalomyelitis, experimental allergic, 171–172
Experimental allergic encephalomyelitis, 171–172
Expressed sequence tags, sequencing, 9, 11–12
Extracellular processes, as therapeutic targets, 71–72

F(ab) fragments, phage display, 83, 85–88
Fibrinopeptide A7-16, mimetic to, 123–124

GenBank, 21
Gene expression, targeting and control, 54–63
protein–protein interactions, 55, 62
temporal or transient fashion, 54
Gene insertion, 41
cellular modalities, 39–41
viral vectors, 41–50
adenoviral, 45–49
other vectors, 49–50
retroviral, 41–45
Gene therapy:
common candidate disease for, 38
industrial applications, 46–49
nonviral systems, 48
polycation systems, 48–49
retroviral systems, 46–48
options, 63
Genetic engineering technology, molecular pathology mapping, 74
Genetic vaccination, 53–54
Gene transfer, by receptor-mediated endocytosis, 51–52
Genome, inheritance and, 7–8
Genome-based therapy, bioinformatics role, 20–24
Genome Sequence Data Base, 21
Genomic mismatch scanning, 18
Genomics, therapeutic, in private sector, 31–32
Glycines, oligo N-substituted, drug discovery, 94–97
Growth factors, binding to membrane receptors, as therapeutic target, 69, 71

Helper T cells, 158
Herpes simplex virus type 1 vectors, 50
HIV protease inhibitors, 145–147
HLA-linked autoimmune disorders, 155
Homologous recombination, 60
Human anti-murine antibody, 114–115
Human genetic maps:
developing and applications, 14–17
Internet access to, 13
Human genome:
gene therapy, 37–64
mapping strategies, 13–20

physical map, 14–17
technological developments, 39
see also Delivery systems; Gene insertion
Human Genome Initiative, 1–13
benefits from, 4
databases, 21
disease diagnosis at molecular level, 4–5
implications of technologies developed for, 5
motivations for, 1–2
Human Genome Project:
databases, 23
data generated per day, 22–23
Human growth hormone receptor, design of potent antagonists, 142–143
Human neutrophil elastase, inhibitors, 88–89
Huntington's disease, 24–25
Hybridization subtraction techniques, 9
Hydantoins, solid-phase synthesis, 80–81
Hypoxanthine phosphoribosyltransferase, abnormal gene for, 8

Immune intervention, 177
immunosuppressive agents, targeting T cell, 177–178
strategies, 181
targeting
accessory molecules, 178–179
cell adhesion molecules, 179
cytokines and related molecules, 180
Immune regulation, 168
Immune response:
important interactions, 158
overview, 154–155
Immune response, re-engineering, 154–182
mechanisms of tolerance induction, 166–169
MHC blocking strategies, 172–175
T-cell vaccination, 176
Th1-Th2 balance shift, 176–177
TCR blocking strategies, 175–176
Immune system:
antigen processing and presentation, 161–166
MHC genes and products, 161–163
organs and cells, 155–159
processing pathways, 164–165
trafficking, 159–161
Immunoglobulins, superfamily complementary determining regions, 121
Immunotherapy, monoclonal antibody engineering, 114, 116–117
Inheritance, genome and, 7–8
Inositol phosphate-glycan, breakdown, 178
Institute for Genomic Research, 31
Interleukins, 160

Kat expression vectors, 47
Knockout mice, 60

Lesch-Nyhan syndrome, 8
Ligand, 68
Lymphocytes, 155–156
activated, 159
Lymphoid organs, 155–159

Major histocompatibility complex, 161–163
blocking strategies, 172–175
molecular structure, 165–166
molecules, 48
Marker, 6
Mimetics, reverse-turn, 121
Molecular diversity, 75–76
methods for creating, 77–79
methods for quantitating, in combinatorial libraries, 104–105
oligonucleotides
randomized libraries, 97–99
RNA ligands, 99–101
peptides, 90–97
chemical synthesis, 91–94
soluble synthetic peptide combinatorial libraries, 90–91
using oligo N-substituted glycines, 94–97
proteins, 83, 85–89
directed evolution, 88–89
phage display of antibody F(ab) fragments, 83, 85–88
small organic molecules, 101–103
Molecular dynamics, 127
Molecular mechanics, classical, drug design, 126–128
Molecular medicine, 37–39
Molecular recognition units, as radioimaging diagnostics, 119
Monoclonal antibodies, 157
molecular recognition units as radioimaging diagnostics, 119
Mouse, linkage map, medical applications, 19–20
Mouse models, generation, 56–62
Cre/Lox system, 61–62
gene switches, 56, 58, 60
knockout mice, 58–60, 62
transgenic mice, 56–57
Multifactorial diseases, 7
Multiple sclerosis, murine model, 171–172
Mutagenesis, codon-based, 85
Mycophenic Mofetil, 178

Natural products screening, 73
Nonviral systems:
gene insertion, 50–54

Nonviral systems: (*Continued*)
cationic liposomes, 50–51
genetic vaccination, 53–54
polycation conjugates, 51–53
industrial applications of human genome, 48
Nuclear magnetic resonance, drug design, 139–141

Oligomeric libraries, 78
Oligonucleotides:
antisense, 97–98
molecular diversity, 97–100
solid-phase synthesis, 94
Organic compounds, screening and identification through peptide tags, 101–102
Organic molecules, small, molecular diversity, 101–103

Parallel synthesis, 82–83
Partial least squares method, 135
Parvovirus vector, 50
Peptide engineering, 120–124
Peptides:
chemical synthesis, 91–94
combinatorial libraries, 88–89
MHC blocking, 173
molecular diversity, 90–97
soluble synthetic combinatorial libraries, 90–91
spatial light-directed peptide synthesis, 91–92
tags, 101–102
therapy, EAE model, 173–174
Peptidomimetics:
engineering, 120–124
retrosynthetic synthesis of reverse-turn, 121–122
Peptoids, 94–97
solid-phase synthesis, 94–95
Peripheral tolerance, 166–168
Phage display, proteins and antibodies, 85–87
Pharmacophores, 78–79
definition, 125, 129–134
development from series of delta-selective opioid cyclic peptide, 130
mapping, drug design, 128–129
Phospholipase, as therapeutic target, 72
Polycation conjugates, gene insertion, 51–53
Polycation systems, industrial applications of human genome, 48–49
Polygenic diseases, 7
Positional cloning, 6
Propriocidal regulation, 172
Protease, specificity, combinatorial exploration, 93
Proteins:
engineering, 110–120
knowledge-based modeling, 112–113
understanding of protein structure, function, and stability, 111–112
molecular diversity, 83, 85–89
Provirus, 42
Purine nucleoside phosphorylase, inhibitors, 144

Quantitative structure-activity relationships, 135–136
analysis of ACE inhibitor series, 131–132
drug design, 129, 133, 135–137
Quantum mechanical methods, drug design, 126

Radioimaging diagnostics, molecular recognition units as, 119
Receptor-effector interactions, pharmacological definitions, 73
Receptors, 68
Recombinant retrovirus, propagation, 42–44
Recombination, 157
homologous, 60
Renin/angiotensin system, as therapeutic target, 71
Representational difference analysis, 18
Retinal pigment epithelial cells, gene therapy, 40–41
Retinosis pigmentosa, 27–29
Retroviral vectors:
gene insertion, 41–45
problems associated with, 42
industrial applications of human genome, 46–48
Retrovirus:
life cycle, 41–43
recombinant, propagation, 42, 44
Rheumatoid arthritis, Vβ14 expression, 175
Ribozymes, 100–101
RNA, molecular diversity of ligands, 99–101
Rous sarcoma virus vector, 49

Sequence-tagged site map, 2–3
Serial analysis of gene expression, 18–19
Sex chromosomes, 7–8
Short tandem repeat polymorphism, class markers, 15
Shotgun sequencing, 22
Signal transduction pathways, 69
rationale for targeting, 69
Split synthesis method, 80, 82–85
Statistical methods, 133–135
Structure-based drug design, *see* Drug design
Synthetic unrandomization of randomized fragments, 98

Systematic evolution of ligands by exponential enrichment, 99–100

T cells:
immunosuppressive agents targeting, 177–178
vaccination, 176
T-cells receptors:
alpha/beta, 156
blocking strategies, 175–176
genes, immune system
Therapeutic products:
novel, from biotechnology, xiv–xv
in phase III clinical trials, xvi–xvii
Therapeutic targets, in living cell, xviii–xix
Th1-Th2 balance, 168–169
changing, 176–177
Thymidilate synthase, inhibitors, 144–145
TIGR, 31
TNF-&alpha, antibodies against, 180
Tolerance, induction:
with antigen, 169, 171–172
mechanisms, 166–169
central, 166
immune regulation/suppression, 168
peripheral, 166–168
Th1-Th2 balances, 168–169
Transgenic mice, generation, 56–57
Tumor, transduced, human genome, 47–48

Very large-scale immobilized polymer synthesis, 91–94
Virally directed enzyme prodrug therapy, 46–47
Vitronectin receptor, as therapeutic target, 72

X-ray crystallography, drug design, 138–139

Yeast artificial chromosome close, 4–5